ADVANCE PRAISE

Liechty skillfully tells the inspirational story of alternative development of hydropower infrastructure in Nepal. Over the course of five decades, Odd Hoftun and Balaram Pradhan did not just build power projects, through the Butwal Power Company, but laid the foundation for a vibrant domestic power sector in Nepal with thousands of competent engineers and technicians, over a hundred active independent power producers, and dozens of construction and manufacturing companies and engineering consultancy firms. This book is a must-read for development practitioners and aid agencies interested in critically examining how to get the best returns from international development assistance.
—**Bikash Pandey**, Director of Clean Energy, Winrock International, USA

Nepal's hydropower sector has been a flame that has, over the last seven decades, both illuminated her development space and attracted exuberant moths (national and international) to their doom. In this book, Liechty describes the success of a remarkably innovative moth whose story has not been properly told before. It is that of an alternative pathway whose relevance has only grown as the conventional development landscape faces the end of the Age of Aid and the rise of one of climate change concerns. It must be read, not just by Nepali hydrocrats but by all development professionals in international agencies as well.
—**Dipak Gyawali**, Nepal Academy of Science and Technology, and former Minister of Water Resources of Nepal

This book tells a remarkable story of the growth of local hydropower development in Nepal. It formally starts in 1963 with an agreement between the Nepal government and a Christian organization, the United Mission to Nepal, for the establishment of the Institute of Technology and Industrial Development in Butwal. The driving force in this was Odd Hoftun, an experienced applied engineer from Norway. It was a time when development economists were discussing the pros and cons of quick turnkey infrastructural projects with external consultants, and so on, or going a slower route and building a local engineering capability to construct and manage institutions in the hydropower sector. Odd Hoftun and his Nepali colleagues persisted with the second route and this book describes what happened from the early days to the present time. The book focuses on the story of the private construction company, the Butwal Power Company (BPC), which emerged from the early initiatives. This history includes public and private takeovers, major changes in government policy, local and foreign investors, and much, much more. *What Went Right* tells this story in careful detail and shows the influence of BPC on Nepal's current hydro sector. The book is a "page turner" as Liechty takes us from one set of major events in construction and politics to the next … and the influence of BPC's legacy is still unfolding in major ways.

—**Stephen Biggs**, SOAS, University of London

If asked, most of us would likely explain Nepal's many development challenges as stemming from its status as a flailing minnow sandwiched between the behemoths India and China, as an enduring Shangri-La mired in its ancient cultural and familial heritage, or as the inexorable legacy of harsh colonial imposition and contemporary neoliberal hubris. *What Went Right* provides a more nuanced and compelling alternative: Nepalis have long sought enhanced well-being, but on their own terms and by finding locally legitimate solutions to their particular development problems, especially as it pertains to harnessing energy from water. Mark Liechty carefully highlights how difficult, fraught, and contingent such a strategy is, and how long it takes to fully consolidate itself, but in so doing demonstrates how respectful partnerships, dogged persistence, and sustained grassroots improvisation can succeed where so many other top-down technical approaches have stumbled.

—**Michael Woolcock**, World Bank and Harvard University

WHAT WENT RIGHT

What Went Right explores why Nepal's hydropower sector is one of the country's few development success stories. Unlike almost every other "developing" country, in Nepal local firms design and build complex hydropower facilities using Nepali engineers, builders, components, and labor. Nepal has largely avoided the trap whereby most poor countries are forced to accept energy infrastructure projects that are foreign designed, funded, and built—typically resulting in debt, dependency, and unsustainability.

This book traces the half-century history of the Butwal Power Company and the anti-establishment development logic of its founder, Odd Hoftun. A pioneering Norwegian engineer, development worker, and missionary, Hoftun insisted that, if Nepal was to create a modern national economy, Nepalis must develop technical skills needed to break the cycle of poverty, a view that led Hoftun to promote Nepali-driven hydropower development as the key to Nepal's industrial future. Counter to prevailing development logics (then and now), Hoftun insisted that all aspects of hydropower development (design, construction, manufacturing, maintenance) be done in Nepal, by Nepalis.

The book traces the struggle between two competing development paradigms: one that emphasizes gradual national human capacity building (at the expense of speed and efficiency) and another that emphasizes rapid, large-scale infrastructure building (at the risk of unsustainability and dependency). At stake is whether what passes for "development" primarily benefits the countries in which it occurs, or the banks, corporations, and other investors that finance capital-intensive projects. *What Went Right* brings a vision for sustainable development into vigorous conversation with other development strategies that have proven, repeatedly, to be less productive.

Mark Liechty is professor of Anthropology and History at the University of Illinois at Chicago. Liechty has studied Nepali history and culture for over three decades. He is the author or editor of five books on modern Nepal and middle-class culture and is a founding co-editor of the journal *Studies in Nepali History and Society*. He has co-edited, with Michael Hutt and Stefanie Lotter, *Epicentre to Aftermath: Rebuilding and Remembering in the Wake of Nepal's Earthquakes*, published by the Press in 2021.

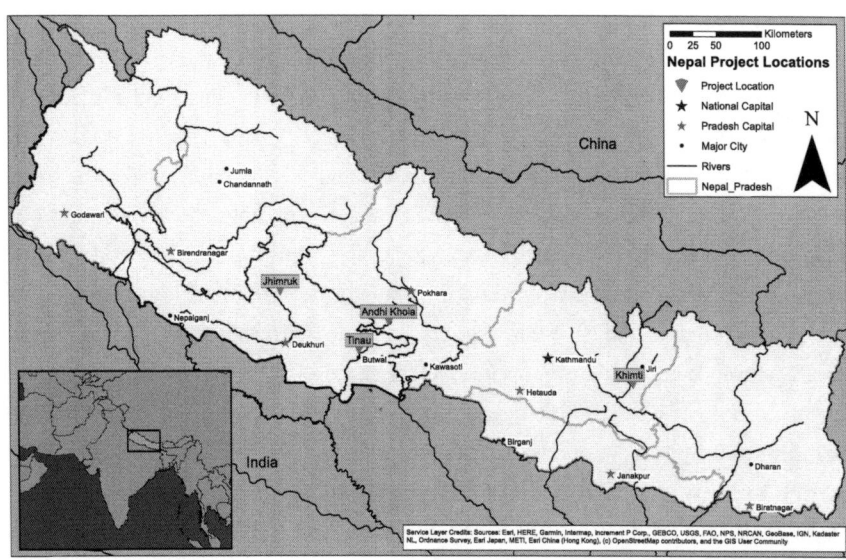

Nepal Project Locations

Source: Holly Aslinger.
Note: Map not to scale and does not represent authentic international boundaries.

WHAT WENT RIGHT

Sustainability versus Dependence in
Nepal's Hydropower Development

MARK LIECHTY

CAMBRIDGE
UNIVERSITY PRESS

University Printing House, Cambridge CB2 8BS, United Kingdom

One Liberty Plaza, 20th Floor, New York, NY 10006, USA

477 Williamstown Road, Port Melbourne, vic 3207, Australia

314 to 321, 3rd Floor, Plot No.3, Splendor Forum, Jasola District Centre,
New Delhi 110025, India

103 Penang Road, #05–06/07, Visioncrest Commercial, Singapore 238467

Cambridge University Press is part of the University of Cambridge.

It furthers the University's mission by disseminating knowledge in the pursuit of education, learning and research at the highest international levels of excellence.

www.cambridge.org
Information on this title: www.cambridge.org/9781316514900

© Mark Liechty 2022

This publication is in copyright. Subject to statutory exception
and to the provisions of relevant collective licensing agreements,
no reproduction of any part may take place without the written
permission of Cambridge University Press.

First published 2022

Printed in India by Thomson Press India Ltd.

A catalogue record for this publication is available from the British Library

ISBN 978-1-316-51490-0 Hardback

ISBN 978-1-009-08896-1 Paperback

Cambridge University Press has no responsibility for the persistence or accuracy of URLs for external or third-party internet websites referred to in this publication, and does not guarantee that any content on such websites is, or will remain, accurate or appropriate.

For
Odd and Tullis Hoftun
and
Balaram Pradhan

Human capacity building does not work by directive from above. Nor does it fit well under crisis management, where the time perspective is too short. It is more like having the faith and courage to *go and plant a seed*. Then adding water and some fertilizer, but otherwise standing aside, letting the process take its time—and simply watching it grow. Both people and institutions are living organisms that must be allowed to find their own way and shape.

> —Odd Hoftun (from a presentation to the African Ministerial Conference on Hydropower and Sustainable Development, Johannesburg, South Africa, March 2006; italics added)

CONTENTS

Preface XI
Acknowledgments XV
List of Abbreviations XVII

1. A CORPORATE VISION
Business as Development Philosophy
1

2. THE BUTWAL TECHNICAL INSTITUTE, TINAU, AND THE ORIGINS OF THE BUTWAL POWER COMPANY
28

3. ANDHI KHOLA
59

4. JHIMRUK
83

5. THE "GREAT UPHEAVAL"
Khimti and the Limits of the Hoftun Hydropower Vision
113

6. MELAMCHI AND THE RUSH TO PRIVATIZATION
148

7. PRIVATIZATION
The Long Haul
194

8. THE NEW BPC
Cultures in Conflict
230

9. CONCLUSION
From Seed, to Plant, to Seed
270

Bibliography 299
Index 304

PREFACE

Nepal's hydropower sector is one of the country's few development success stories. Nepal has engineering, construction, and manufacturing capabilities that drive a thriving national market in hydropower installation and production. While virtually every other comparable country in the "developing world" relies on foreign expertise and equipment, Nepali companies independently produce everything from simple water-powered grain mills to large-scale hydropower components, and from sophisticated hydraulic modeling services to skilled hydro-engineering and design services. Specialized Nepali construction firms draw on skilled Nepali labor to build large hydroelectric power plants of up to around 100 megawatts (MW) (Karki 2017: 122–123). Nepal has even begun to export these skills, services, and components to other parts of Asia and Africa (R. S. Shrestha et al. 2018: 6). Developments in Nepal's hydropower sector are still well behind those of advanced industrial economies, but they are also well ahead of other "least developed countries." As a whole, Nepal has no other industrial sector that even comes close to the success of its hydropower industry.

This book examines the history of Nepal's hydropower sector to ask why it is the conspicuous exception to the rule of Nepal's woeful underdevelopment. The answer, I argue, lies in the story of the Butwal Power Company (BPC) and the antiestablishment development logic of its founder Odd Hoftun, a pioneering Norwegian development worker, missionary, and engineer. From the early 1960s onward, Hoftun insisted that Nepalis should develop technical skills needed to thrive in a modernizing society, a view that eventually led Hoftun to promote hydropower development as the means to literally power Nepal's industrial future. Counter to the then (and now) prevailing logic, Hoftun insisted that, to the extent possible, hydroelectric design, construction, and equipment should be locally sourced—even if it was, initially, crude and inefficient. Self-sufficiency and sustainability could only come, Hoftun argued, if every aspect of hydropower development could be done in Nepal, by Nepalis. In the face of ridicule from everyone—from foreign and Nepali peers to big international development agencies—Hoftun started small but, over the course of half a century, worked with Nepalis and other foreigners to establish a technical training institute and, gradually, a family of interlocking companies focused on hydrological design and engineering, equipment manufacturing, deep-mountain tunneling, and project installation. Starting with a tiny 50-kilowatt (kW) project in the 1960s and advancing through successively larger and more complex projects, by the 2000s Hoftun's now independent and Nepali-owned companies, and many subsidiary spin-offs from them, had emerged as the backbone of a robust indigenous hydropower sector able to compete successfully in bidding for projects around Nepal and beyond. Although Hoftun is a key figure, my aim is less to portray him as a hero than to systematically examine how his approach unlocked Nepali human potential in ways that other approaches to development have not.

Typically anthropologists and historians engage "development" in order to critique it. In a world where "First World"–driven "Third World" development initiatives have almost universally failed, in spite of the trillions of dollars spent, there is good reason to critically examine the institutional, ideological, and economic factors that both doom development initiatives to failure and guarantee their perpetuation. An entire interdisciplinary subfield—Critical Development Studies—has now intensively explored this problem (Veltmeyer and Bowles 2018; Veltmeyer and Wise 2018).

Much less often critically examined are the few bright spots on the global development landscape. My aim is certainly not to hold up Nepal's hydropower sector as some spotless paragon of development success but simply to examine how and why it managed to largely overcome the global development odds stacked against it. To be sure the Nepal story is rife with tensions inherent in any market-based approach (profiteering, corruption, perpetuation of social inequality, and so on). Still, by the standards of global capitalism, Nepal's hydropower experience is a success and one of the few ways that Nepal participates in the global economy aside from as an impoverished exporter of cheap manpower. By focusing on *what went right* instead of (or in addition to) what went wrong, I see this book as a useful contribution to ongoing debates over international development, foreign aid, and development philosophy.[1]

Reminiscing about his career in Nepal, in 2015 Odd Hoftun—a sly smile on his face—told me, "There is one thing that, in retrospect, I have regretted … and that's that we didn't have a propaganda department! We never thought about that." Rather than letting his work (hopefully) speak for itself, Hoftun wished that they had been more intentional in sharing his vision with the Nepali public broadly and opinion leaders in particular. His ideas of gradually increasing human capacity in a slow but sustainable way caught on among likeminded Nepalis connected with BPC and its many projects. But beyond that small circle—in the face of those who supported big plans, using big money, in a big hurry—Hoftun's vision was easily overlooked. Most Nepali officials, foreign aid experts, and Nepali professionals "were thinking in the opposite way. And they are *still* thinking the same way! To build from the bottom up is not their way. They want to come in in the modern way and just get things done, their way," said Hoftun.

In this book I certainly do not aim to provide one-sided "propaganda." Rather, this study represents a chance to lay out a particular development vision to examine its strengths and weaknesses. Given that Hoftun's development vision has arguably borne rich fruit in Nepal's otherwise relatively barren development landscape, it is high time to bring Hoftun's vision into vigorous conversation with other development strategies that have proven, repeatedly, to be less productive.

---

[1] The story of Nepal's hydropower development industry, and its unusual success, has been largely ignored—though this is beginning to change. A recent study of aid, technology, and development in Nepal (Gyawali, Thompson, and Verweij 2017) is highly critical of most donor-driven "development" initiatives in Nepal since 1951 but points to the Hoftun–BPC small hydropower development approach ("the BPC model") as one of Nepal's few development success stories. Even more pointedly, the authors of a recent paper point to what they call a "substantive paradox": "despite tremendous gains in local technical and institutional knowledge and even manufacturing in the hydropower arena, why is more positive recognition not given to the growing and strong hydro sector in Nepal?" (R. S. Shrestha et al. 2018: 6). This book aims to at least partially resolve this paradox by providing not just "positive recognition" but a detailed account of one crucial facet of Nepal's hydropower success story.

ACKNOWLEDGMENTS

As with most scholarly productions, this book is the result of contributions made by many, many people. First and foremost is Odd Hoftun, who graciously and patiently allowed me to interview him over an intensive two-week period in 2015. He also gave me unlimited access to documents ranging from handwritten notes, email logs, and news clippings to minutes, diaries (in Norwegian!), and annual reports. Odd also read and extensively commented on the book manuscript. While the book's framing and interpretations are mine, and I take full responsibility for them, I have tried to remain true to the task of analyzing Odd's development philosophy and documenting its evolution and consequences. Even if the final version does not always represent all of his personal priorities, I hope that this book clearly reflects the great respect that I have for Odd Hoftun, his work, and his legacy.

Another key contributor was Balaram Pradhan. Although I was only able to spend a relatively short time with him (before his untimely death in 2016), I felt that I got to know him somewhat just from the countless times that he appeared in the archival sources that I used. I also credit his suggestions for helping me clarify the objectives, priorities, and audience for this book.

I am also deeply grateful to the twenty-five or so Nepali and expat hydropower and development experts who kindly agreed to be interviewed about their experiences with, and perspectives on, the Butwal Power Company and its legacies. Almost all of them are quoted (or their views otherwise presented) in the text. In order to respect their privacy I have chosen not to acknowledge these people by name yet I want to be clear that the book would not have been possible without their cooperation and I am sincerely thankful for their contributions.

I would like to specially acknowledge the people who generously read and commented on various drafts of the manuscript: Stephen Biggs, Dipak Gyawali, Odd Hoftun, Marjorie and Russel Liechty, Paul Myers, as well as five reviewers who have requested anonymity. I am responsible for the remaining errors of fact and interpretation.

Finally, sincere thanks to the following: Martin Chautari for access to Hoftun's archived United Mission to Nepal (UMN) files; Naomi Liechty for companionship on several site visits in Nepal and for providing photographs for this volume; Holly Aslinger for preparing the map and timeline; Bikas Rauniar for the cover photo; Tullis Hoftun, Leif-Egil Lorum, Tor Møgedal, and Peter Svalheim for generous assistance, hospitality, and sharing of insights during my visits to Norway; Peter Lockwood, Duane Poppe, and Dan Spare for generously accompanying me on site visits in Nepal; and, for miscellaneous acts of assistance and support, to Erik Hoftun, Laura Hostetler, Dale Nafzinger, Biraj Pradhan, Tara Pradhan, Tara Lal Shrestha, and Mark Windsor.

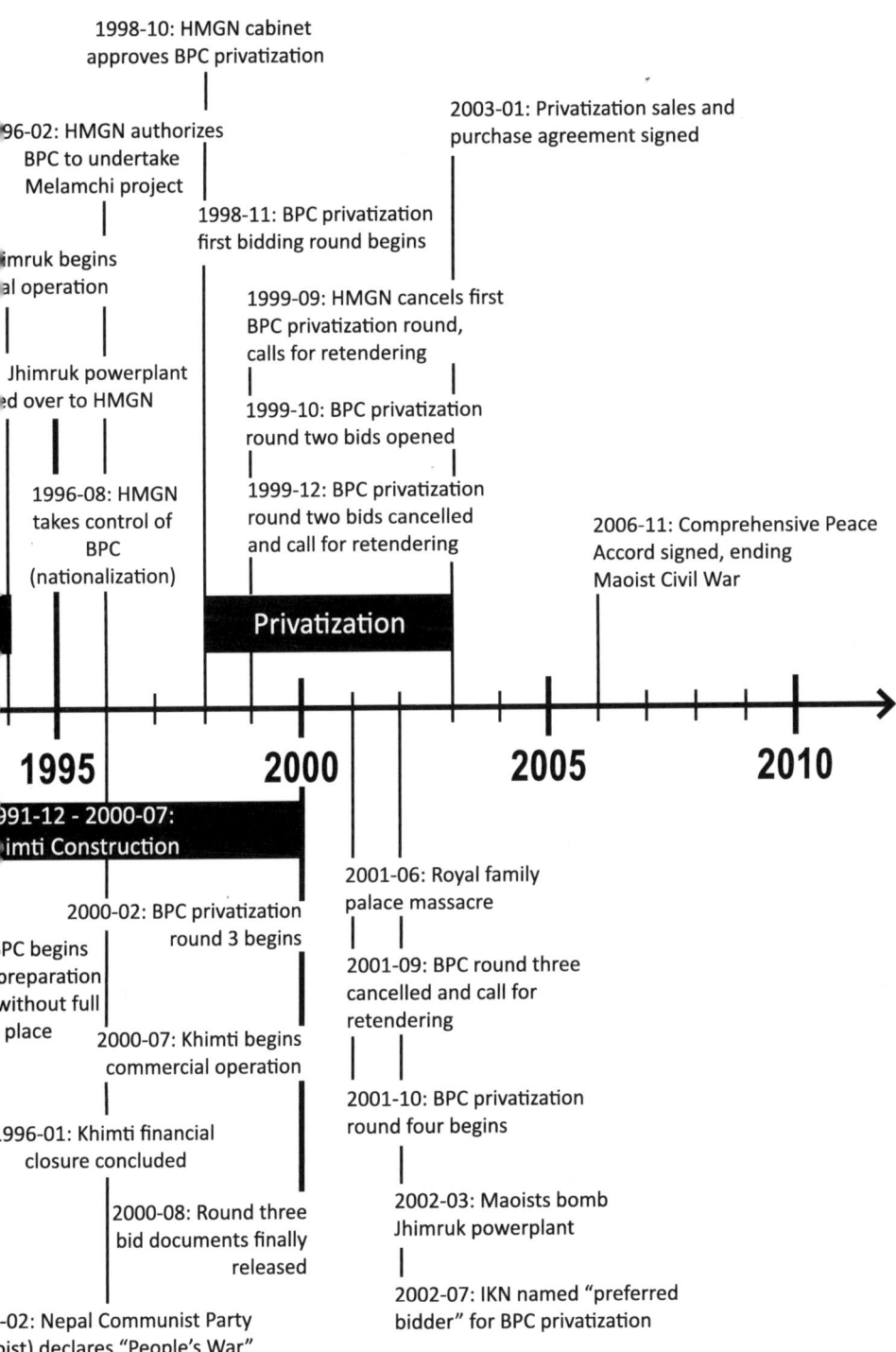

ABBREVIATIONS

ADB	Asian Development Bank (headquartered in Manila)
AKWUA	Andhi Khola Water Users Association
BEW	Butwal Engineering Works
BPC	Butwal Power Company
BTI	Butwal Technical Institute
CDO	chief district officer
DCS	Development and Consulting Services (a division of UMN)
EDC	Electricity Development Center
EIA	environmental impact assessment
FITTA	Foreign Investment and Technology Transfer Act
GON	government of Nepal
HMGN	His Majesty's Government, Nepal
HPL	Himal Power Limited (corporate owners of the Khimti project)
IFC	International Finance Corporation (commercial finance wing of the World Bank)
IKN	Interkraft Nepal (Norwegian and Nepali group organized to invest in BPC)
INTERKRAFT	Norwegian hydropower utility consortium
IMF	International Monetary Fund
IPC	Independent Power Corporation (British commercial power developer)
IPCN	Independent Power Corporation Nepal
IPP	independent power producer

IPPN	Independent Power Producers of Nepal (trade group)
IRD	integrated rural development
ITID	Institute of Technology and Industrial Development
JICA	Japan International Cooperation Agency
JIDC	Jhimruk Industrial Development Center Pvt. Ltd.
JRP	Jhimruk Rehabilitation Project
KEC	Khimti Environment and Community (mitigation unit of HPL)
KHL	Khudi Hydropower Limited
KV	kilovolt
KW	kilowatt
MDS	Melamchi Diversion Scheme
MOWR	Ministry of Water Resources
MP	member of parliament
MPMP	Multi-Purpose Melamchi Project
MVA	megavolt-ampere
MW	megawatt
NEA	Nepal Electricity Authority
NGO	nongovernmental organization
NHE	Nepal Hydro and Electric Pvt. Ltd. (hydropower equipment manufacturer)
NIDC	Nepal Industrial Development Corporation
NOK	Norwegian krone
NORAD	Norwegian Agency for Development Cooperation (the Norwegian state development agency)
NPR	Nepal rupees
OPEC	Organization of Petroleum Exporting Countries
PPA	power purchase agreement
PPM	parts per million
REE	rural electrification entities
SEL	Shangri-La Energy Limited (Nepali investors in, and eventual owners of, BPC)
SLC	school leaving certificate
SPA	sales and purchase agreement

STATKRAFT	a Norwegian state-owned hydroelectric development corporation
UML	Unified Marxist–Leninist (Party)
UMN	United Mission to Nepal
UNDP	United Nations Development Program
USAID	United States Agency for International Development
USD	United States dollar
WKV	Wasserkraft Volk

1

A CORPORATE VISION

Business as Development Philosophy

Two commonly cited facts about Nepal stand in uncomfortable tension. On the one hand, Nepal's thousands of rivers cascading down thousands of meters of Himalayan slopes have the *potential* to generate phenomenal amounts of clean, cheap, renewable hydroelectric power that could, in turn, fuel a vibrant, sustainable industrial and consumer economy, making Nepal the envy of its neighbors far and wide. But, on the other hand, today Nepal's main export in the global economy is human labor: mainly cheap, unskilled, and easily exploitable. Unable to find jobs at home, millions of Nepalis—10 percent or more of the country's population—work in low-wage occupations mainly in India, Southeast Asia, and the Gulf States. As one Nepali expert asked, "Why is there a lack of employment? Because there is lack of industrialization. And why don't we have industrialization? Because we lack necessary power for the purpose" (Shrestha 2011). As such, to say that Nepal is blessed with hydropower resources is something like a cruel joke: instead of exporting power and industrial goods, Nepal exports young men and women.

As any Nepali knows, this situation is hardly new. Nepal's hills and mountains have historically stood as food-deficit, population-surplus zones providing labor to surrounding lowland areas, just like many other poor mountainous regions around the world.[1] Part of Nepal's national identity revolves around its history of foreign-bound *lahure* mercenary labor (made

---

[1] Compare, for example, Switzerland's long-standing export of mercenary troops to Italy (think "Swiss Guards" at the Vatican) or the workers from Lesotho in South African factories.

famous by the British as "Gurkha" troops) and the figure of the brave, *bir-bahadur* Nepali hill man (Onta 1996).

A century ago, Norway was in a similar position. A cold, mountainous region with a short growing season, little arable land, and few (known) natural resources,[2] Norway was among the poorest countries in Western Europe.[3] But one resource that it did have was snow-fed mountain rivers. By the early twentieth century, tapping those rivers for hydroelectric power was a national priority. Norway began its gradual rise in hydropower production, industrialization, and standard of living. Today, 95 percent of Norway's electricity comes from hydroelectric generation, Norwegians heat their homes mainly with electricity, and have the highest per capita rate of electric vehicle use in the world, while Norwegian companies are global leaders in electromechanical equipment for hydropower generation.[4]

Odd Hoftun *lived* that transformation. Because his father was an electrical engineer who built and ran a small, community-owned hydel-generating plant in a rural, upland region of central Norway, Hoftun grew up literally surrounded by the challenges and rewards of turning moving water into electricity and then distributing it as a service to consumers.

> My father was the manager of a local electric utility involving generation as well as distribution, and I grew up in a power plant the same way as a farmer's son grows up on a farm. I learned that the business of an electric utility is to serve the public. It means being available any time during day or night—when things go wrong and need to be fixed. It is by nature a monopoly situation—very different from running an industry that produces goods for sale in the open market where competition immediately sorts out what is sound and what is fake. Therefore the utility has to be

---

[2] Since then, North Sea oil has proven to be a huge boon to the Norwegian economy, but those resources were only tapped beginning in the 1960s.

[3] Norway was also forced to export labor—in the form of immigrants to North America and seamen. Norwegian ships and crews provided transport services around the world, a legacy that lives on today.

[4] For example, Kvaerner (now known as "Rainpower") still holds patented proprietary designs for high-head, highly efficient turbines.

service minded and do its utmost to deliver what the public requires. (Hoftun 2004)

As a child and young adult, Hoftun watched as increased power production across Norway led to greater employment opportunities, more stable and sustainable livelihoods, and—in the broadest terms—increased national prosperity. Power production was a key to unleashing human potential and promoting national dignity. But surrounded by technology and technical problems, Hoftun also knew that without highly skilled human resources—from machinists to engineers to physicists—all the flowing water in the world would not produce electrical power. Water is only a resource when it is harnessed, and that requires major investments in both human and financial capital.

As a young man, one of Hoftun's main priorities was developing his own human capacity. Inspired by his father, at university Hoftun studied electrical engineering, but he wanted to equip himself with other skills as well. As Hoftun remembered,

> It was just after the war [WW II] and … Norway was very much worn down and behind in every aspect. Still there was great enthusiasm during those years after the war, at least until the coup in Czechoslovakia. That period was a fantastic time for young people. It was inspiring. To do something that was meaningful, that was the goal.

Hoftun saw university as a chance to prepare not just for a job but for a life in which he could "do something meaningful." As such, he took extra courses in economics, accounting, and construction—skills that prepared him to be a generalist, not just a specialist. "That was in my thinking from almost when I started there," Hoftun recalled.[5]

The "great enthusiasm," optimism, and service ethic that Hoftun describes emerged, for him, at the intersection of both secular and religious ideals. Like the rest of his generation, Hoftun had an intense desire to see his homeland

---

[5] Any quotation not accompanied by a formal citation is taken from interviews conducted by the author.

uplifted: service to the nation—to advance the common good—was an inspiring life goal. Like many Norwegians, Hoftun embraced a basic social-democratic political ethic that took shape in a relatively progressive, quasi-socialist, welfare state that publicly upheld the values of equality and justice. Although secular, many of these state values resonated with the values of Norway's (until very recently) state-affiliated reformed Lutheran (Protestant Christian) Church. Hoftun was raised in this denomination and, as a young adult, chose to anchor his identity and life's work in this faith tradition. The product of this specifically Norwegian state and Christian orientation, and trained as a generalist engineer, Odd Hoftun—along with his new bride, Tullis—left for Nepal as a missionary with the United Mission to Nepal (UMN) in 1958.

To understand Hoftun's work in Nepal, it is important to recognize the strongly convergent state/secular and church/religious morality that Hoftun, like many other northern Europeans, was socialized into. Where many Asians, North Americans, and others might see a sharp clash between state and religious values, Hoftun's Norway provided a model of how secular ethics could comfortably overlap with religious commitments. Social justice, service to others, equality, honesty, hard work—these and many other values could just as easily be accommodated within secular and religious worldviews.[6] In

---

[6] One thing that struck me in my conversations with Hoftun was how his Christianity coexisted alongside a very real sympathy for leftist, even radical leftist, politics. As hinted at in the block quote earlier, in the postwar years, many of Hoftun's generation in Europe had real hopes for some of the Socialist experiments underway around the world—as in Czechoslovakia, until it was snuffed out in a Stalinist coup. In another conversation I noted China's slide toward capitalist gerontocracy, to which Hoftun replied wistfully, "And yet we had such high hopes for China when it began." Similarly in Nepal, while most of his fellow missionaries would have been quick to equate any form of communist or Marxist ideology with godless evil, Hoftun recognized leftist Nepalis as at least potential collaborators. Balaram Pradhan, Hoftun's longtime friend and close associate in Nepal, was an outspoken member of Nepal's Unified Marxist–Leninist (UML) Party, a movement that Hoftun matter-of-factly described to me as "basically a social-democratic party." A party that many Westerners (especially missionaries) would have instinctively condemned, Hoftun saw as similar to many mainstream secular progressive parties in Europe. Similarly, when Nepal's Maoists laid down their arms and won the popular election in 2008, Hoftun was optimistic and

Nepal, Hoftun remained deeply, even stubbornly, committed to a particular vision of ethical practice. But, unlike many of his missionary colleagues from other national and faith backgrounds, Hoftun was comfortable with framing his ethics in secular terms. Many, many Nepalis told me that while Hoftun never tried to hide his Christian faith, neither did he ever try to impose it on Nepalis or imply that Nepalis were incapable of behaving ethically without becoming Christian. I believe this fact was fundamental to Hoftun's ability to attract dozens, even hundreds, of like-minded Nepalis who broadly shared his ethical vision and who, together, materialized it across half a century of shared work in Nepal.

TANSEN

In 1954 the United Mission to Nepal (UMN) established a hospital in an old rented building in Tansen bazaar in Nepal's Palpa district. When it became clear that the hospital needed its own larger, specialized medical facility, UMN acquired land a few kilometers east of the town and invited a young missionary and engineer from Norway, Odd Hoftun, to come and build it. When Odd and Tullis arrived in 1958, they found conditions in Nepal to be both strikingly foreign and uncannily familiar. Observing poor, landless Nepali day laborers crouched by the roadside breaking rocks with hammers (for use in road construction), the Hoftuns saw not just poverty but also visions of the past when they had seen poor Norwegians doing exactly the same thing only a few decades earlier (Svalheim 2015: 7). In a strange way, the Hoftuns had traveled to the other side of the world only to find a version of home. Nepal and Norway were vastly different but also surprisingly similar in terms of the challenges they faced and the resources they had to confront those challenges.

---

looked forward to working with people like Hisila Yami and Baburam Bhattarai, who he believed would share at least a basic social justice ethic. My point is not that Hoftun is some kind of closet Marxist or Communist sympathizer, but rather that he is able to see in leftist politics the potential for pragmatic collaboration based on shared ethical values.

When it came to building a hospital—in addition to having limited practical experience in construction—Hoftun faced two big problems: a shortage of money and a shortage of skilled labor. Technically, Hoftun knew what he was supposed to do and how to do it. But with very little (and sporadically available) money to work with, steel and concrete imported from India were out of the question, and even bricks and timber were expensive and in short supply. Nothing happened quickly, and patience was one of the key lessons that Hoftun learned in his four years at Tansen. Another lesson was the need for improvisation and flexibility. Without money and standard materials, Hoftun had to innovate on the job, a task he worked on in collaboration with another breed of UMN workers known as Pax men— young American Mennonite[7] men who, expressing their pacifist convictions, volunteered to do service work overseas in return for exemption from military conscription in the United States. Almost all of these young volunteers had grown up on farms in rural USA where they had learned to solve problems with few resources. "They were *not* doing things in the latest fashion!" Hoftun recalls. "So, they got onto the practical problems that we had when we started from absolutely scratch in the beginning, and that worked very well." These Pax men were masters of, and for Hoftun models for, how to "make do"—how to get things done even without the right tool or the right materials. Their "can do" and "make do" abilities helped Hoftun develop his own ideas about how things could and should be done in Nepal (cf. Horst 2018: 177–178).

If lack of money was one problem, lack of local skilled labor was another. Hoftun needed Nepali craftsmen to help build this hospital, but there were virtually none available. Many people were looking for work, but the kind of worker Hoftun wanted was not there. Hoftun quickly recognized that if he wanted skilled labor, then he and others were going to have to train people themselves. So they did. The Tansen hospital got built using a variety of innovative, cost-cutting, locally sourced techniques. Hoftun watched as two things arose: a hospital and a core group of semiskilled Nepali masons,

---

[7] Mennonites are a Protestant Christian denomination who believe that followers of Jesus should, as Jesus taught, practice nonviolence, serve others, and "love your enemies."

woodworkers, electricians, blacksmiths, and so on.[8] What's more, once the hospital was done, the Nepali craftsmen took their skills and used them as the basis for their careers. It dawned on Hoftun that he had solved two problems: building a hospital and giving a group of people the skills to stand on their own feet. On-the-job training may not be the most efficient way to simply get a job done (like building a hospital), but it was a very efficient way of producing skilled labor that would make the next building job much more efficient.

It was from this realization that Hoftun began to contemplate a way to institutionalize the kind of skills training and job creation that had happened haphazardly on the Tansen hospital building site. Hoftun began to raise money for a trade school and approached the Nepal government with the idea, only to be turned down. The government did not want more Christian "mission schools." After a furlough and further thought, Hoftun came back to Nepal and pitched his ideas for a training institution, not to education officials (as he had done previously) but to the Ministry of Industry. And this time his proposal to promote industrial skills came with an ultimatum: allow the plan to go ahead in Nepal, or it will be taken to India. Once framed in terms of industry, not education, in 1963 the Nepal government agreed—on the condition that the new training institute be located in the Tarai, where transportation would be less challenging. Everyone agreed on Butwal because of its geographic centrality: on the border between the hills and the Tarai, it was also near the intersection of the old road from India to Pokhara and the new East–West Highway then under construction.

What follows is the story of how Hoftun brought together UMN and Ministry of Industry officials to establish an Institute of Technology and Industrial Development (ITID). Although the ITID agreement had a larger scope,[9] the institute was soon known popularly as the Butwal Technical

---

[8] Tullis Hoftun contributed by teaching some of the most promising young craftsmen reading, writing, and basic mathematics.

[9] In addition to a technical training institute, the ITID agreement called for a design and engineering division, though this later function was not institutionalized until 1972 with the founding of the Development and Consulting Services (DCS). See Chapter 2.

Institute, or BTI. To provide electrical power for its own workshops, the institute built a small hydropower plant on the Tinau River (just above Butwal) that eventually grew into a 1-megawatt (MW) facility. In order to own and operate the plant and its distribution network, Hoftun and UMN registered the Butwal Power Company Ltd. (BPC) in 1965, in partnership with the Nepal government, which became a minority shareholder. By the time the Tinau project was finally completed in 1978, BTI was flourishing but Hoftun was beginning to formulate a new vision for further work centered on hydropower. From his experience in Norway, he knew the powerful "multiplier effect" that energy brought to any economy. With energy came jobs and industry, and with jobs came money circulating through the community, which, in turn, created more jobs. If, like Norway, Nepal could turn its moving water into power, that power would form the basis for an independent national economy. And as a Norwegian, familiar with Norway's history of conquest and exploitation by foreign states, Hoftun saw hydropower as a key element in fledgling Nepal's struggles to assert sovereignty and economic independence vis-à-vis its giant neighbors, India and China.

Hoftun also believed that for hydropower development to sustainably contribute to Nepal's industrialization and national independence (economic and political), it had to be firmly in Nepali hands. Even if well intentioned, power plants engineered, built, and maintained by foreigners would leave Nepal *dependent* on foreigners. Nepal needed the skilled labor to take a power project from concept to completion. As he had learned while building the Tansen hospital, Hoftun knew that creating skilled labor through on-the-job training was time consuming and inefficient. (If your goal is to simply get a hospital or power plant built fast, it is much more efficient to use money to import supplies and already skilled labor.) But he also knew that once those Nepali skills were in place, they would form the basis for greater and greater accomplishments. Looking back at his early career in Nepal, Hoftun laughs when noting that he stumbled upon ideas that would later become development mantras: appropriate technology, sustainability, bottom-up development, technology transfer, and, above all, capacity building. Capacity building is what would ultimately break what Hoftun often referred to as the "vicious circle": without experience, no jobs; and without jobs, no experience. On-the-job training would provide jobs and experience, on the basis of which one was qualified for further jobs. Whether for an individual or a company or a nation, capacity building was the key to building competence, sustainability, and independence. As a Nepali former BPC administrator explained to me, BPC existed not to produce money, electricity, or even power plants but to produce "trained Nepalis."

BPC became the means by which to implement a vision of increased power production, job growth, improved standards of living, and national independence. After the 1 MW Tinau project, between 1981 and 1991 BPC and its subsidiaries built the 5 MW Andhi Khola power project. From 1988 to 1994 they built the 12 MW Jhimruk project. And from 1993 to 2000 BPC and partners built the 60 MW Khimti project. In the meantime, BPC was first nationalized (in 1995) and then privatized (in 2003), at which point it became the Nepali-owned hydropower developer and distributor that it is today.

Looking at BPC's trajectory it is impossible not to make an organic analogy: the company and its accomplishments were like a growing organism. In this book's epigraph, Hoftun himself compares human capacity building with planting a seed. But in my interviews with him, Hoftun was insistent that, while he may have known that he was planting a seed, he had no preconceived notions of what that seed would grow into.

> Although my background was in electric power business, that was not at all in my mind, either for coming to Nepal or for doing things after we started. Rather, it grew out of [unforeseen circumstances]. There was perhaps some vision of rivers and the power potential they had in terms of industrial development and building up jobs for people and the economy in general. That was the overall concern or aim.... [But] there was no idea about one project building up to the next. That just came out of the situation: we have to train people and get going on our own. We were just trying to figure out how to do things.

"The way things progressed was more or less accidental or coincidental, not planned long in advance," Hoftun continued. "There is nothing in BPC history that was really planned. It has grown up out of the environment of development in Nepal." This book examines how the seed that Hoftun planted grew, with the help of many Nepali hands, in the distinctive political, cultural, and historical environment of modern Nepal.

But before telling that story (beginning in Chapter 2), it is essential to explore Odd Hoftun's development philosophy: its characteristics, how it evolved in Nepal, how Nepalis working with Hoftun understood it, and why many Nepalis recognized Hoftun's vision as a version of their own dreams for Nepal. It is this convergence, I believe, that accounts for the successes BPC achieved and, perhaps, for some of its challenges today.

DEVELOPMENT PHILOSOPHY

Many have noted that the Butwal Power Company (BPC) was an unprecedented organization in Nepal. For a church-related development organization (the United Mission to Nepal [UMN]) to support Hoftun's vision for promoting private enterprise by establishing a private commercial corporation (BPC) was as pioneering as it was controversial. And because the government of Nepal was from the beginning a shareholder in the company, BPC had folded into its corporate DNA three distinct, and in some ways conflicting, institutional logics. First was that of UMN: although church based, UMN was basically what would later be called a "nongovernmental organization" (NGO) dedicated to serving the disadvantaged and operating on a nonprofit, grant- and donation-funded basis. The second institutional logic was that of the government: a bureaucratic administrative body that exists to (hopefully) advance the national good, paid for by taxation (and, in the case of Nepal, international aid). And third was the logic of business and the market: profit-driven, competitive, and inherently risky. In effect, these three represent three epistemological stances: donor-driven service, prioritizing the national interest, and market-derived profit.

In some ways the BPC story—from its founding under UMN, to its nationalization and state ownership, to its eventual full privatization—seems to proceed sequentially through these three institutional forms (what I will call nonprofit, government, and business). But it is not that easy. In fact, all three sets of institutional logics and values have been present in BPC from the beginning and, to some extent, remain to this day. Analytically, it is important to ask the relative weight given to any one of them at a given moment in BPC history. And, even more pointedly, we need to trace the tensions and contradiction between these three epistemological impulses: to seek equitable benefit sharing, to advance the national interest, and to profit.

Some of the basic questions I asked Odd Hoftun (and many others who had been involved in BPC) were: Why the corporate or business model? If the goal was to promote capacity building in the hydroelectric sector, why not form an NGO or foundation that would provide this service? Or why not work with the government to establish state-sponsored training programs? Why try to root your vision in the competitive capitalist market with its "survival of the fittest" logic? How can you preserve the values of service and human dignity in a corporation where the "bottom line" (profit) is usually the ultimate value? Or, as many of Hoftun's UMN colleagues pointedly asked over the years, how can a Christian, and especially *a Christian missionary*, even think about promoting big business that is inherently exploitative of the very people the missionary aims to serve? As one fellow missionary complained of Hoftun's UMN-backed enterprises:

> I cannot see the compatibility between a mission and the owner of a large industrial complex. The goals clash! Companies happen to be associated with big money (profit motive) and a position of power. How to reconcile this with our missionary calling of lowness (stepping down) and servanthood? Why do we want to run the risk of being misunderstood?[10]

Hoftun endured decades of criticism from missionaries for whom, as he said, "'company' is a bad word."

If the market-based corporate model was so bad, why not work to create a government-run public enterprise? In many ways BPC was, and still is, a "public utility." Its power distribution business requires it (in theory) to be "service minded." For Hoftun, the nationally owned, service-oriented public utility was an ideal. At least initially, he was happy to hand over the power

---

[10] Hoftun archival material at Martin Chautari, UMN Economic Development Board records, a letter to Wynn Flaten, EICS, from Ben van Wije, re: EID Long Term Plans, November 7, 1990.

projects BPC developed into government hands. But Hoftun quickly learned that public ownership did not necessarily translate into service for the public good. Whether due to lack of resources, lack of administrative efficiency in a young bureaucratic state, or due to sheer incompetence and/or corruption, Hoftun soon saw that placing hydropower infrastructure into government hands was an invitation for disaster.

If a government-owned enterprise was not the ideal platform on which to develop his human capacity-building ideas, why not start an NGO or foundation? For Hoftun, the private, nonprofit model is good to the extent that it allows its management to maintain control over the ethical principles that guide the organization. But the fatal flaw of the NGO model was, for Hoftun, its dependency on donor funding. Furthermore, NGOs tend to reproduce the logic of donor dependence in their own programs, which often treat aid recipients as charity cases who, in turn, become dependent rather than self-sustaining. But in terms of efficiency, looking around at the development world in Nepal, Hoftun saw the main danger of NGO donor dependency to be not a lack of resources but *an overabundance*. "To have too much money is like poison," Hoftun told me. Rather than fostering institutional budget management, problem solving, and independence, donor money promotes waste and inefficiency. "Too much money just destroys the will to make do with whatever is possible and keep the costs down. I learned that stuff while we built the hospital in Tansen. It was a good lesson."

Ultimately, Hoftun embraced the corporate, market, business model of institution building for a combination of practical[11] and personal reasons. Corporate institutions minimized some of the flaws of the other approaches, but they also offered the means by which to pursue a vision of ethical business practice that was rooted in Hoftun's specifically Norwegian Christian upbringing. Hoftun described to me a tradition or ethos within the Norwegian state church that promotes "Protestant frugal values but at the same time is not anti-business. [A Christian should] just work and

---

[11] As discussed further in Chapter 2, Hoftun also noted that the corporate model, with its relatively standardized legal framework and rules, was an internationally recognized form that facilitated transactions with other businesses and foreign donor agencies.

earn profit, but according to certain standards."[12] Far from being a place that Christians should flee, the market *could be*—if one lived "according to certain standards"—an ethical space. While many (including the UMN colleague quoted earlier) took the idea of "ethical capitalism" to be inherently contradictory, in the Norwegian context there was a much more idealistic view of the market and its possibilities. Far from being necessarily antithetical to them, the market was a place in which to cultivate Christian values.

Part of Hoftun's belief in the ethical market grew out of his experiences in Nepal. Already in Tansen it was clear to Hoftun that because of their vastly greater power (in terms of wealth, education, access, mobility, and so on), missionaries faced a huge barrier in their efforts to relate with Nepalis. To overcome this barrier, one of the main temptations was to *give* resources to Nepalis. But, Hoftun found, rather than overcoming differences, charity only amplifies them. Charity creates givers and receivers, lenders and debtors, donors and beggars, independent and dependent. Hoftun described how he and his wife learned this lesson the hard way. Lending money to Nepalis "was an almost automatic barrier-building thing. So we just learned what we should not do."

By contrast, Hoftun argued, when one deals with another on a commercial, business basis, the power differential is not erased, but at least it is minimized. Business implies a contract between legal equals, both of whom receive something of relatively equal value in an exchange and both parties are (ideally or in theory) free to accept or reject the contract.[13] Hoftun noted,

> This is one principle that I have stuck to in private life and in what I have been in charge of in Nepal. When dealing with people about money and

---

[12] Svalheim's biography examines the uniquely Norwegian Christian roots of Hoftun's sense of business ethics in more detail, especially the influence of Hans Nielsen Hauge.

[13] Critical theorists have often rejected the idea of the market contract as a freely entered leveling device. Marx focused specifically on the labor contract, arguing that so-called "free laborers" (destitute people with nothing left to exchange but their own bodies) were coerced into *unequal* "exchanges," for which the formally legal labor contract served as an ideological justification for exploitation and the basis for the extraction of "surplus value."

things like that, we are commercial. We want something in exchange for something else. So we are equals in that sense.... We don't give money. There must be a clear cut line that, *we don't do that*. We do business. And we want to have something in return.

For Hoftun, a market-based business relationship promotes equality and human dignity. When I pressed him on whether he believed that business relations were inherently conducive to equality and equal exchange, Hoftun paused and said, "Well, yes, if it is *honest* business."

This crucial caveat points to the crux of an ethical dilemma that Hoftun struggled with throughout his decades in Nepal. He acknowledged that the business contract *does not* eliminate power differentials between the two parties, a differential that draws into question the voluntary basis of the contract itself. Therefore, for the market to be an ethical space, the parties operating in it must be ethical. And furthermore, it is especially important that the more powerful party in any exchange behave ethically. It is the obligation of the dominant party not to abuse its power. *If* used ethically, money and power can be a force for good in the business marketplace.

This "if" helps to explain one of Hoftun's lifelong business obsessions: corporate ownership and control. Specifically, when derived from ethical business practice, financial profit is a good thing. *Any* business needs profit to be self-sustaining (in the face of competition), and that necessity promotes discipline, frugality, efficiency, resourcefulness, innovation, and so on—all qualities that Hoftun saw as ethically positive and absolute keys to business success. But the question becomes, *what does one do* with the profits, and *who gets to decide*? "Profit is such a dangerous thing," Hoftun told me. "How do you spend that profit? Profit is a way you measure whether [a company] is sustainable, able to carry on by itself. But sustainable for what purpose?" Ironically, profits are both the key to corporate sustainability and a source of grave danger.

For Hoftun, profits can be good or bad. Profits are bad when used by individuals or groups to enrich themselves at the expense of others. Profits are good when they are invested in the growth and development of the corporate group. Just as a woman selling vegetables in the bazaar would be condemned for hording her profits and not using them to nurture her children,[14] the owner of a corporation should be condemned for reaping personal benefit from a collective enterprise, rather than using that profit to enhance the well-being and capacity of the collective. Quite simply, profit is to be plowed back into the company, not skimmed off the top. The difference between these two outcomes—corporate greed or corporate responsibility—is a question of ownership or who gets to control the decision-making. For Hoftun, "The nature of a company depends on the owner. Corporations aren't bad, leadership is."

This fact points to one of the primary reasons that Hoftun embraced the private, market-based, corporate model in which to pursue his goal of promoting human capacity building in Nepal. Hoftun believed that, at its most basic level, the market is an ethical space of free and fair exchange. The tradesmen trained on the Tansen hospital site, and later at the Butwal Technical Institute, entered a simple marketplace in which they exchanged skilled labor for a fair wage. However, problems arise when *corporations* enter the marketplace. Rather than relatively equal individuals engaged in exchange, when corporations act in the market, corporate leadership has the potential to *use individuals* (employees, laborers) for the personal gain of the corporate owners. Or, as Hoftun put it, "It's mainly at the leadership or owners level where these conflicts come up." It was precisely because ownership guaranteed control over leadership that Hoftun chose the corporate model as the best means by which to pursue his vision. Hoftun explained,

> I have no doubts about the company, the limited company model. It is dominating most business in the world. It is a good model. It puts finances under clear control. So for the sake of control … it is best.

For decades, and even now, Hoftun worked tenaciously to maintain control, or at least influence, over the corporations he founded in Nepal in hopes of keeping those enterprises on the straight and narrow path of ethical behavior. As long as he and like-minded people were in control, Hoftun was confident

---

[14] Or a man who drinks his wages on the way home from work.

that he could curb the corrupting influence of money, engage in ethical business practice, and spread that ethical corporate culture to others.

In fact, it was the possibility of modeling ethical business practices—and perhaps even transforming business culture in Nepal—that Hoftun saw as one of his primary goals. Hoftun was (and is) quick to condemn standard business practice in Nepal as "corrupt," "dirty," and "rotten." Bribes, kickbacks, fraudulent accounting, tax evasion, secret deals—Nepalis view these practices as, at best, social aspects of the local business culture or, at worst, the unavoidable demands of doing business in Nepal. But, for Hoftun, they represented an unethical, irrational, greedy corruption of the marketplace. Especially in the face of missionary colleagues who were quick to condemn *all business practice* as evil, Hoftun resolutely advocated ethical, even Christian, business practice. Reading through UMN archival materials that document UMN's management and oversight of programs Hoftun was involved in, a *frequently* recurring theme is the epic "struggle" to do ethical business in Nepal.[15] In one telling debate from 1983, one of Hoftun's UMN critics asked, "Does it fall within the purpose of the UMN to be involved in enterprises … which operate according to the rules of competition and can therefore give little attention to social and human development?" Hoftun replied pointedly,

> It is our belief that, instead of trying to avoid business enterprises, Christians actually have a duty to demonstrate that business enterprises

---

[15] The language used by UMNers to describe this issue is remarkable. For example, in one discussion of "Guidelines on Business Ethics," a committee resolved to:

> 1) recognize the constraints placed upon our work by prevailing practices and attitudes, 2) combat these practices and attitudes as much as possible in order to strengthen the distinctive Christian witness already established in related industrial institutions, 3) continue to confront and struggle with these difficult issues. (Hoftun archival material at Martin Chautari, Industrial Development Planning and Management Committee, minutes of the second meeting held on 15 September, 1983)

> Here, Nepali business culture *constrains* ethical practice. It is something to be *combatted, confronted, and struggled* against as in some epic conflict between good and evil.

can be run and a fair return obtained, and at the same time function as a means of social and human development. In a country like Nepal where business enterprises are of growing importance and may be the only hope of employment for many in the future, it is all the more important for UMN to try to demonstrate in a humble way how this can be achieved with integrity and dignity for all. In particular, we do not see profit as the sole or even the most important goal for a business enterprise. Profit is certainly a pre-condition for survival, but we believe that a viable enterprise will be able to get beyond this stage of survival to one in which the fruits of ingenuity and hard work can be shared by all in an equitable way.[16]

In conversations with me, Hoftun spoke of defending his work to colleagues engaged in health care, education, agricultural development, and so on. Like them, he aimed to model Christian values to the people he encountered, mainly in government and business. Half-jokingly, he told me, "Just like running a hospital or a school, it is also a mission field to work with government people, and rotten businesspeople too!" And more seriously, "Dealing with this rot has been a constant struggle. I would say it has been one of the toughest things I've done in Nepal." Even his most sympathetic supporters recognize something quixotic in Hoftun's crusade against corruption. But even those less sympathetic to Hoftun's crusade acknowledge that he never failed to live up to his own strict standards. As one of Hoftun's Nepali colleagues told me, "Other foreigners might bend [compromise their ethics], but not Hoftun."

Before concluding this discussion of how and why Hoftun chose the corporate, market-based approach to institution building, it is important to point out some of the anomalies, or even contradictions, that play out in BPC's corporate history. Rather than being clearly distinct from the nonprofit, NGO or state-owned models, BPC was often a somewhat bizarre amalgamation of all of them. For example, while BPC was established as a private corporation,

---

[16] Hoftun archival material at Martin Chautari, Industrial Development Planning and Management Committee, minutes of the first meeting held on February 6–7, 1983, app. 5.

the Nepal government was an important shareholder from the beginning.[17] What's more, because BPC had (and still has) a monopoly over power distribution rights in certain parts of Nepal, it functions as a "public utility." Therefore, BPC has to be "public minded" and able to "combine openness and public service with commercial efficiency and private sector dedication and drive" (Hoftun 2004). BPC is a kind of public/private hybrid.

But the logic of the donor-driven NGO world also plays an important role in BPC's past. Hoftun is quick to acknowledge that for its first thirty years, BPC and its corporate offshoots were almost completely dependent on grant and donor funding. Before it started the 60 MW Khimti project in the 1990s, all of BPC's power projects were grant funded, mostly from Norad, the Norwegian state development agency. Hoftun recognized the anomaly of a private sector company subsisting on grants—especially in light of the pointed criticisms he had for NGOs that were grant dependent. "My own philosophy was that grant money should be used as if it had been obtained on commercial terms," writes Hoftun (2004). In other words, even if grants did not require BPC to pay interest or dividends, Hoftun insisted that donor money be used very frugally with all profits to be invested in "sound expansion" of the business.

BPC's donor funding points to one of the prime ironies of Hoftun's decision to follow a private-sector, market-based approach: the reason that BPC could not rely on the market for its support is that for decades the market did not exist. Quite simply, the market for hydropower services, and even electrical power itself, hardly existed. Therefore, *Hoftun had to essentially create the market demand, and simultaneously simulate market conditions,* for BPC and its spin-off companies. Unless and until there was market-driven demand for hydropower investment, hydropower engineering services, hydropower construction services, and hydropower electromechanical equipment, Hoftun would have to mimic the conditions under which such demand operated so that in time, when demand finally arose, his companies would be ready. In the process, some accused Hoftun of manipulating the market and artificially dictating commercial relations between the companies. From Hoftun's point of view, all the companies were part of a larger strategy. And, more to the

---

[17] A point discussed in more detail in the next chapter.

point, he felt that the different companies (representing different sectors of the hydropower development economy) needed to *learn* how to work both independently and interdependently. They needed to learn how to operate efficiently, profitably, and sustainably; how to negotiate legal contracts; how to meet deadlines—in short, how to operate in the market even if the market was not there.

Hoftun's corporations were like test-tube babies or exotic animals raised in captivity. They were not born of the market with the necessary instincts ingrained. Rather, they were birthed artificially and had to be socialized into the "real world" of the market. The hydropower development sector has been likened to an "ecosystem" in which many different parts (investment, design, construction, equipment, distribution) interact symbiotically. What Hoftun had to do was set this ecosystem in motion. He knew from experience elsewhere how the ecosystem worked, but in Nepal it had to be carefully coaxed into place such that (1) the different pieces actually existed, and (2) the different pieces learned how to work independently but together in a *system*. Eventually the parts must learn to work together and, simultaneously, in their own interest which is, ultimately, the interest of the system. Every part of the ecosystem should thrive and multiply, but none can thrive without all the other parts of the system. Viewed in this light, what Hoftun was able to accomplish was little short of heroic—creating both supply and demand for private, market-based hydropower services.

BPC'S CORPORATE ETHOS

But as impressive as these accomplishments are, Odd Hoftun would be the first to acknowledge that whatever successes he had in Nepal were the result of countless people, mostly Nepalis but also many other expat volunteers, who helped him refine and actuate his vision. Without them, little would have happened. Hoftun was an inspiration and guide to many. But the fundamental question is how and why Nepalis embraced his vision. The answer, I believe, lies in the way that Hoftun's personal values and goals resonated deeply with like-minded Nepalis who, with him, made Butwal Power Company (BPC) history.

Many former and current BPC staff whom I spoke with stressed that it was easy for them to meet Odd Hoftun in the secular ethical middle ground between their Hindu or Buddhist and his Christian religious orientations. Couched in secular terms, Hoftun's ethical principles were entirely compatible with their own. Many Nepalis also pointed out that, compared with many other UMNers, Hoftun never pushed his religion on Nepalis. Hoftun's longtime close associate Balaram Pradhan said how many UMNers were more interested in proselytizing than in serving others.

> But Odd was never like that. He had his vision for something else, and not for this [promoting conversion]. And we used to talk. I said, "I don't think there is only one [religion]. I don't know why you people are interested to convert this and that." And he was agreeing with me, saying that, "Yes, everybody should read Bible, Gita, Koran, or whatever, and let the people feel free to follow whatever he likes." So we were of that group.

Pradhan went on to describe how Hoftun was never shy about praying or following other Christian practices while around him. Or, when Pradhan was visiting him in Norway, how Hoftun would invite him to go to church, which Pradhan enjoyed for the "calm and peace." Both men were religiously committed, and both could respect that in the other. This combination of firm ethical belief along with religious tolerance attracted many like-minded Nepalis into Hoftun's camp.

Part of this view came through in Hoftun's management style. While he might have been tolerant, Hoftun could also be judgmental, in good and not-so-good ways. On the one hand, he was quick to condemn those whose business practices he disagreed with, often labeling them with terms (such as "dirty" or "rotten," mentioned earlier) that were morally loaded, uncompromising, and disparaging. But, on the other hand, Hoftun was equally quick to identify what he saw as good qualities in others and to reward them. Both Nepalis and foreigners often commented to me about how Hoftun's meritocratic management style was very un-Nepali. He preferred organizational structures that were relatively nonhierarchical, in which everyone could be involved in decision-making and, in particular, where everyone was invested in the values, strategies, and goals of the organization. He also had zero tolerance for anyone using their position for personal advancement or any kind of shady business. For Hoftun, *loyalty* was a key word and a basic metric of a person's value to the organization. "Being dedicated to the kind of work we were involved in—the training and industrial development—that is what we looked for," said Hoftun. Hard work, collective orientation, innovative thinking, loyalty—these qualities led to greater responsibility within the organization, with younger people sometimes advancing more quickly than those with seniority. Although it violated Nepali bureaucratic norms, Hoftun's emphasis on quality of service (over length of service) created a dynamic corporate culture that was exciting and rewarding to work in.

Almost all of the former and current Nepali BPC staffers (who had worked with Hoftun) I spoke with stressed how the combination of Hoftun's commitment to service, honesty, and openness, along with his relatively egalitarian, meritocratic management style, promoted an energizing sense of team membership. One person described weekly BPC staff meetings in the 1970s and 1980s where people from various parts and levels of the organization would meet and "share openly, hiding nothing because there was nothing to hide." He described a feeling of calm, focus, camaraderie, and high morale that was possible because everything was transparent and there were no unethical things going on. Those above were not holding secrets from those below. No one was exploiting their position for personal gain. People worked collectively to solve problems in a way that gave everyone a sense of both shared responsibility and shared accomplishment. Perhaps most tellingly, many Nepalis spoke of how, during their BPC years, they felt that they were not working *for* a company but *as* a company; not *for* the bosses but *with* a variety of teammates. "There was no difference between owners and employees," said one person. "We were *all* employees during our time. Now there are employers and employees. That is the difference that I see in one sentence."

Working in BPC not only gave people a sense of collective mission but also a target for that mission: an opportunity to serve their country and advance its collective destiny. "BPC gave us good opportunities to develop our careers and to serve our society," said one person. Another remembered the intense pride and satisfaction he derived from contributing to the Andhi Khola irrigation scheme, which transformed a dry mountainside into a lush green agricultural zone. Many noted Hoftun's fierce "love for Nepal" both because it mirrored their own and because it was a vote of confidence that stood in contrast to the opinions of most other foreigners. Several people told me that Odd Hoftun was one of the greatest Nepali nationalists they had ever met! "I don't know how it is now," another former staffer reflected, "but every BPC staff during that time [the earlier decades], feel proud to say that 'I was a BPC employee!' I feel proud to be part of BPC history because BPC existed to serve Nepal, not to serve some owners." He went on to explain how today Nepal is able to build complex hydropower projects that neighboring Bhutan could not even contemplate. "I know the value of these things. There is a sort of feeling that you have for your nation. You want to stand on your own. That feeling, it doesn't come in a day."

Hoftun may have been the spark that lit the fire, but it is important to see how many Nepalis (and other foreigners) added the fuel that allowed BPC and its accomplishments to grow exponentially. What's more, although Hoftun continued to play important roles, BPC as an institution soon took on a life of its own with Nepalis, in many cases, leading the way in modeling the company's values and mission. Many of the Nepali staffers I met did not actually have much direct contact with Odd Hoftun. If anything, most had more contact with Nepali administrators like Balaram Pradhan and it was Pradhan's leadership that many pointed to as inspirational. One of the Nepali engineers that Pradhan hired remembered how

> I didn't know much about him from before. But when we started to work together, he didn't show selfishness in a single hour—not in a single minute!—while talking or eating together or whatever. I will say this openly, that this helped me to continue for this long. If I hadn't had a person like Balaram leading, along with his whole team.... I mean, when you have good people in leadership and you form a team, *then those other team members also turn out to be good*! That's the good thing that I've found. If you have good thinking and good objectives, and no self-interest, then people do not dare to blame you or take advantage of the situation.

His point is that Pradhan helped to establish a culture of openness and honesty that was valued by all. "And this whole thing we learned from BPC.... If you are fair and truthful, then nobody can point to you and say, 'Oh, you have done wrong.' We had a clean and transparent mechanism to work with and that helped me." BPC's values might have their roots in Hoftun's vision, but they were turned into *institutional values* by the many people, mainly Nepalis, who collectively built the institution by *choosing* to support it.

I emphasize choice because virtually all of the Nepali professionals who joined BPC (and its corporate spin-offs) could have chosen differently. Almost all the former and current Nepali engineers I spoke with described how, as young graduates, they were expected to enter government service. Government jobs were seen as prestigious, respectable, and secure. And because they promised a relatively easy life of white-collar office employment in Kathmandu, most young Nepali engineers took government jobs in roads, irrigation, electricity, mining, and so on. But others chose to work for BPC even if the private sector was seen as very risky. Some chose BPC because they saw the government as being corrupt in a way that would force them to go against their own ethical principles. One person described how

> I was very much committed to clean business principles. At that time, when I started [my career] there were plenty of opportunities with the government but it was a very corrupt profession. So I actually jumped into that [the private sector and BPC], taking a very great risk. But that was not the risk [I was worried about]. Going to the government sector was more of a risk for me.

Related to this was a sense that BPC was involved in "positive change." Government service may have looked prestigious to outsiders, but insiders reported a feeling of lethargy and institutional inertia (with much tea drinking and newspaper reading). By contrast, BPC's "working culture" was dynamic and progressive with an emphasis on using technology innovatively to bring about positive social change.

All the engineers I spoke with said that another major attraction of BPC, compared to government jobs, was its strong emphasis on training and skill building. Government jobs often required even engineers to engage in administrative, planning, liaison, and other kinds of nontechnical work. But joining BPC, which usually had several projects under design and construction, meant that a young engineer would immediately enter an active work environment with countless opportunities to acquire and build practical skills. Many commented that their engineering competence and confidence developed far more rapidly working with BPC than they would have in a government job.

What made Hoftun's vision of BPC as a training, capacity-building institution possible was the company's access to a large pool of experienced international engineering professionals who volunteered their services through UMN. These people worked with recent Nepali engineering graduates who may have had theoretical knowledge but virtually no practical experience. When I asked one former BPC engineer what he learned after joining BPC, he replied,

> I should say most everything! I had two years of technical education, mostly on theory. So *everything* I learned from Andhi Khola! We might have heard the word "tunnel." But there we *constructed* tunnels. We might have heard the word "tailrace" but there we turned that into reality in construction and operation of the tunnel. For all technical things, it was a practical field for us—learning as well as doing.

Many commented on the quality of the learning environment that BPC offered—one that focused on teamwork and group accomplishment rather than hierarchy and competition. "It was especially the people I value the most," reflected one person.

> Because those people, I mean UMN had expats from almost twenty countries at that time, so many countries including USA, Britain, Norway, Sweden, Holland.... And they were very nice people and they groomed us because *they were supervisors but they worked as teachers*. So that gave us good opportunity to gain experience to work and give some output to the extent possible, and also to *learn*, every day. And to learn things at an international standard.
>
> And it's not only me. Many people who started [with BPC], they are now spread all over Nepal. In every project we see BPC people. And not

only people with technical or academic qualifications, but also foremen, technicians. (Italics added)

Others remembered the frank and constructive relationships they had with their supervisors and teachers. "When you did something wrong, they would tell you openly. But not in a threatening way. Every mistake was a chance to learn, to get better. We became very close friends." Summing up his BPC years, another person told me:

> These UMN expatriate engineers, I *loved* to work with them. They were always willing to help, willing to teach. As a fresh engineer I lacked many things. But I could learn from them. They always encouraged me. The working environment was *very good*.

But the working environment could also be very challenging: choosing to work with BPC usually meant choosing to work on project sites in rural Nepal, far from the middle-class comforts of Kathmandu that most of the young Nepali engineers were accustomed to. Most BPC work sites followed UMN protocol by which all staff—expats and Nepalis—were expected to live in local housing, eat local food, and intentionally try to harmonize their lifestyles with local standards. "For me, initially, all that was so new that I had problems adjusting," said one person. "But later I got used to it." One engineer told me of his traumatic first assignment with BPC to help construct tunnels in far eastern Nepal. A recent graduate in civil engineering, he had read a few pages on tunneling in a textbook but never dreamed of focusing on, *and working in*, tunnels. Arriving at the construction site, he remembered, "I never imagined that such a remote place existed. It was a terrible place, full of leaches. I felt like I was in some kind of adventure movie, off in some remote part of Africa. That's what it felt like!" Asked to go into the test tunnel, he panicked and for three days sat on a boulder in the river, trying to ward off the leaches.

> For three days I wouldn't go into the tunnel. I sat on my boulder thinking, "OK, this is how Nepal is. I might have to *live this way*." But I gradually

started to learn about what tunneling is. I mean, it's one thing to learn by textbook. But we had some reference books at the site so I was going through those and I gradually got the knack of the tunneling.

Rather than quitting and heading back to Kathmandu for a government office job, he decided to accept the challenges and "live this way." Today, he is one of Nepal's most skilled and experienced tunneling engineers.

CONCLUSION

From its founding, the Butwal Power Company (BPC) was an unusual corporate entity that—under the principled, idealistic leadership of Odd Hoftun, Balaram Pradhan, and others—held a range of sometimes conflicting forces together in fruitful tension. BPC embraced a paradoxical set of potentially incongruous values:

- corporate profit and public service
- ethics and business
- frugality and innovation
- modern technology and appropriate technology
- efficiency/profitability and tedious training and low-tech approaches
- aid dependence to promote market independence
- realism and idealism
- self/career advancement and national advancement
- strict moral values and religious tolerance
- power and power sharing
- strict corporate leadership and team participation with a sense of shared ownership
- a skills gap between expat and Nepali employees, and an emphasis on equality and meritocracy
- working *as* a company, not *for* a company

If, as Hoftun insisted, "corporations aren't bad, leadership is," then the key to maintaining this delicate but dynamic and satisfying balance of corporate drive, ethical practice, and employee empowerment was committed, unwavering leadership. By institutionalizing an ethic of hard work, persistence, professional growth, goal-oriented teamwork, and service to a larger cause, BPC offered Nepalis a career path where the challenges were great, but so were the rewards. Hoftun, Pradhan, and others led by championing and modeling an ethical vision. As an institutional culture, that ethic then attracted Nepalis who shared its vision, helped build the institution, and, ultimately, transmitted its values to new settings.

2

THE BUTWAL TECHNICAL INSTITUTE, TINAU, AND THE ORIGINS OF THE BUTWAL POWER COMPANY

Odd Hoftun, a few other expats, and a crew of increasingly skilled Nepali foremen and construction workers built the United Mission to Nepal (UMN) hospital in Tansen between 1958 and 1963. But long before the building was finished Hoftun had already begun laying the groundwork for a new project that would institutionalize the model of skills training, capacity building, and job creation that he had experimented with at Tansen. Already by the early 1960s plans for a UMN-supported trade school were beginning to take shape—a controversial idea for those who thought that missionaries should only be involved in human services like education and health. Recognizing that such a school would require electricity to operate, on the long walks between Pokhara and Tansen (before motorable roads arrived) Hoftun had already identified the Andhi Khola Valley as a promising building site because of the area's power-generating potential. But when he approached the Ministry of Industry for permission to start a training institute, the minister in charge (himself from the Tarai) requested that the school be located in Butwal, then a small, dusty bazaar town at the foot of the mountains below Tansen. The Tinau River that tumbled down a steep, narrow, rocky gorge just above Butwal was not ideal but had some hydropower potential. A poor road linked Butwal with the Indian border, and footpaths led into the hills to Tansen, Pokhara, and all the way to Tibet. But a proposed East–West Highway on the Tarai would, someday, make Butwal a crossroads. For UMN this foray into industrial training and development was a completely new departure in mission activity but, in spite of some resistance, Hoftun was given the go-ahead.

UMN signed a formal agreement with the government of Nepal on November 7, 1963, according to which UMN would build, staff, and manage an "Institute of Technology and Industrial Development" (ITID) and the government would deed a large parcel of land—about 4 hectares of level ground along the main north–south road, with further land for living quarters extending up into the adjacent hills. Popular usage soon shortened the name to "Butwal Technical Institute" (BTI). The agreement also gave BTI import and tax accommodations, allowing it to bring equipment and materials into Nepal. The original agreement stipulated that UMN would hand over BTI to the government after fifteen years.

Hoftun and others drew up plans for a technical institute in Butwal that would put young men through a four-year training program (based on European models) in which production would be both an end and a means. Trainees would learn production skills while the sale of items produced would go toward making the institute self-sustaining (though not-for-profit). But aside from wood from local forests, materials were virtually nonexistent and had to be brought in from India or abroad. Written into UMN's agreement (largely written by Hoftun) was a commitment not just to give people "on the job" training, but to help identify marketable products and then "encourage and help individuals and firms in the area to take up similar production, and thus to encourage the development of industry" (Møgedal 1983: 4). From the beginning Hoftun envisioned BTI to be a catalyst that would spark private business, employment for skilled labor, and thus, development.

With financing and offers of staff support in place from a German Christian development organization, in early 1964 Director Hoftun, along with German Business Manager Martin Gugeler, started the task of building the technical institute with the help of Nepali craftsmen experienced from their work on the Tansen hospital. Also lending muscle and know-how were more American "Pax men," young Mennonite conscientious objectors who chose voluntary overseas service as an alternative to military conscription back home. In their early twenties, these men had the "can do" spirit and work ethic that came with their farming and/or small-town upbringings. As young volunteers, with no expectation of leadership assignments, they also mixed well with Nepali trainees and workers with whom they often formed strong friendships. After three years these men returned home with valuable experience and transformed worldviews. Many chose development-related careers and some chose careers in Nepal.

While construction was underway, Hoftun returned to Europe to raise further funds, recruit staff, and collect equipment for the technical institute. The Norwegian aid agency Norad gave money as did several religious charities. But most impressive was the outpouring of support from ordinary people, especially in Norway. When people learned about Hoftun's project, through church contacts and media reports, the response was enthusiastic. Donations of new and used equipment came pouring in, people volunteered time to prepare materials for shipping, and a Norwegian shipping company even offered to haul freight to Calcutta, free of charge. (When 170 tons of goods showed up they were shocked but kept true to their word!) From Germany and Norway Hoftun also managed to recruit several seasoned tradesmen to serve as instructors.

With supplies and equipment for the institute on their way, in November 1964 the Hoftun family, along with two other families (twelve people in all), piled into a makeshift camper (a used flat-bed 6-ton Mercedes truck with an old bus body bolted on the back) and headed for Nepal. Six weeks and many adventures later, they pulled into Butwal in time for Christmas.

By early 1965 the institute was beginning to take shape. Goods from Europe were arriving at the Calcutta port. From there Martin Gugeler and assorted Pax men had boxes loaded onto trains to the Nepal border and trucks to Butwal. Though still incomplete, a staff of technical instructors and administrators was also taking shape. These were people who had answered the call for committed Christians who were "professionally on top and still willing to work hard in primitive conditions" (Møgedal 1983: 9). They also had to be willing to look after their young Nepali charges both on the job and after hours as hostel parents. Equipment went into the new buildings under construction but living quarters had to wait: most staff lived in rented rooms in Butwal town. Director Hoftun's office was the old bus body, now on blocks and under a flimsy palm-frond roof to (try to) keep it from turning into an oven in the sun.

With boxes being unpacked and machines put in place, electric power became the need of the day. Staff had a diesel engine and a generator (purchased with Norad money) but had to come up with pulleys, couplings, and brackets using nothing but hand tools. But within a month there was power for a rudimentary mechanical workshop (electric welder, lathe, drill press, and milling machine), among the first of its kind in Nepal. With their trainees, staff instructors began taking in small outside jobs—mainly fabricating or repairing parts. Indian and British crews then constructing the East–West Highway turned to BTI for repair work. Students learned not only technical skills but also how to estimate prices for materials and labor. The institute also began looking for products it could manufacture and sell. This was an opportunity to teach product design, drawing, cost calculation, and production techniques. Soon trainees were producing metal door hardware, ladders, and wheelbarrows along with wooden furniture, doors, and windows—all needed by the institute but also available for display and sale. By the end of 1965, BTI had earned 20,000 Nepali rupees (NPR) for services provided to the community.

THE TINAU PROJECT AND THE BUTWAL POWER COMPANY

As demand for power grew so did the urgency of providing a more sustainable power supply for the Butwal Technical Institute (BTI). From the beginning Hoftun had planned to develop hydropower on the nearby Tinau River above Butwal—initially to power equipment at BTI and eventually to provide power to local consumers. The idea was to start small and gradually expand the generating capacity, both as a way of building up skills and competence and to gradually expand production to meet the hoped-for growing demand. With no existing electricity supply (aside from BTI's expensive diesel generator) or distribution infrastructure, Hoftun could only hope that once power was available, customers would come forward to buy it. Supply had to precede demand.

But more than just faith in the market, building even a small hydel project required significant amounts of money. Already in 1964 Hoftun had applied for, and received, a small grant from Norad (the then recently established Norwegian state development agency) to supply power to BTI. At a cost of 225,000 Norwegian krone (NOK) (c. 31,500 United States dollars [USD]), BTI could import and install a secondhand diesel generator set and a used 50-kilowatt (kW) hydro turbine-generator set, both obtained as gifts in Norway. By April 1965 the diesel generator was up and running, and a few weeks later workers had finished digging a 500-meter canal that would provide water for the provisional hydropower equipment. But when Hoftun returned to the canal site after that summer's monsoon, he found it entirely washed away.[1] It was now clear that, to escape monsoon violence, even small-scale generation was going to have to be built into solid rock through tunnels.

Hoftun had originally envisioned the Tinau hydel project to be one of several divisions within BTI, with BTI trainees producing equipment for the project and eventually getting power in return. But as the construction challenges of Tinau began to grow, so did the need for major financing. Hoftun needed more money from Norad but, as amounts increased, it became more difficult to finance the 50 percent co-investment that Norad required from its partners. They would have to borrow money and attract other investors—all of which, in turn, required some kind of organization that could receive loans and coordinate investments. This was beyond BTI's mission and Hoftun was not interested in government control of the project either. In theory, an NGO could have dealt with Norad. "But," as Hoftun explained to me, "the NGO as such was hardly invented at that time!" Hoftun was negotiating with the Nepal Industrial Development Corporation (NIDC) for a cash loan to help match the Norad grant and it was NIDC's director, Bharat B. Pradhan, who suggested that Hoftun set up a private limited company with the government as a minority shareholder. For raising funds, attracting investment, and keeping strict control on management and finances, Hoftun saw the corporate model as the way to go. And including

---

[1] The Tinau River's monsoon flood rates can reach as high as 1,000 times greater than its flow during the rest of the year.

the government as shareholder was a stroke of genius.[2] On the five-person corporate board the government representative was the official chair, NIDC had a seat, and the United Mission to Nepal (UMN) the other three, with Hoftun as controlling board secretary. Rather than risk potential conflicts with the government, it was better to have them on board in a more or less honorary capacity, with Hoftun and UMN in firm control of the majority.

With UMN's backing, Hoftun finally registered the Butwal Power Company Pvt. Ltd. (BPC) with the Department of Industry on December 29, 1965, with the company's officially stated purpose being to "produce, transmit, distribute, and sell electric power" (Svalheim 2015: 150). In retrospect, having a religious mission organization agree to support a private, commercial, shareholding, development organization appears to have been completely unprecedented—in Nepal or elsewhere. But Hoftun insists that, at the time, they had no sense of being visionary, bold, or pathbreaking. Rather, the birth of BPC was part of a larger organic process; it was just one practical step needed to advance the more immediate goals of developing Tinau and supporting BTI. In his annual report for BTI in 1965 Hoftun expressed frustration but also hope.

> Everything is unfinished and unsatisfactory, and the work we can do is not at all up to the standards we would like. *But through all the frustrations the living organism of a modern industrial enterprise is gradually taking shape.* And just by watching this process people around here learn. Crude and primitive as much of our equipment and work is, it is something people here can copy if they are only willing to learn and work. (Møgedal 1983: 11; italics added)

Here already is Hoftun's metaphor of fostering a "living organism," an image he returned to many times to refer to the complex, interdependent object that he and others nurtured over the years.

---

[2] Of the 50 percent investment to meet the Norad grant, almost all came from the NIDC, some from UMN, and a "nominal amount" from the Nepal government. After the loan was repaid, eventually UMN controlled two-thirds of the share capital and the government one-third (Møgedal 1983: 10).

With BPC officially established, Hoftun could go about raising funds and preparing technical plans for the Tinau power project. Working off of detailed maps prepared by a UMN surveyor, Hoftun planned a three-stage project that would eventually produce a total of just over 1 megawatt (MW) of power. The headrace, drop shaft, powerhouse, and tailrace would all be built into solid rock. Because Nepal had absolutely no legal apparatus in place to regulate the hydro industry, there was no environmental impact study done and no one to be asked for permission to build. "We just took that as part of the deal like we took the water in the Tinau to make power without any restrictions," said Hoftun. By late 1966 BPC had received its first grant or loan money and tunneling could begin.

The problem was that no one really knew how to build a tunnel. Nepalis had been building small-diameter short tunnels for irrigation for a long time. But the Tinau project required a 1.8-meter diameter tunnel almost 2.5 kilometers long. Hoftun consulted tunneling engineers in Norway, read every book on the subject he could find and, with the help of completely inexperienced but willing Nepali workers, began tentatively boring holes into the mountain side. Several fortuitous arrivals helped the process along. One was a Nepali who had worked in road construction, including tunneling, in Kashmir and knew basic drilling and safety procedures. Dhan Bahadur arrived one day looking for a job and immediately became a valuable construction foreman. The other was literally a tourist passing through: Kristian Bockmann, a retired Norwegian engineer with extensive experience in tunneling and mining. Bockmann visited the site and offered some free advice. "He wrote up a two-page recipe for how to handle it," said Hoftun. "And that's what we followed. So that was a gift from above. When you look back, in retrospect it was amazing how things happened and worked out in ways we hadn't planned." Given that no one really knew what to expect or how to proceed, it seems inevitable that nothing could really go "as planned."

Progress was incredibly slow and soon the project was far behind schedule. A pneumatic drilling rig, damaged in transit, turned out to be more trouble than it was worth meaning that virtually all the early tunneling was done by hand, using hammers, chisels, and metal bars for chipping boreholes into the rock. A crew of sixty–eighty men worked simultaneously at four locations via access tunnels, but progress at each tunnel face was only about 1 meter per week. To make matters worse, almost all building supplies came from India, where there were already shortages, rationing, and long waits for things like cement and dynamite. Then in late 1967 Nepal radically devalued its currency against the Indian rupee (Svalheim 2015: 157). Now the project was both seriously behind schedule and over budget. A huge monsoon flood in 1970 delayed things even further. It took four years to dig 400 meters of tunnel. Nevertheless, by mid-December 1970 Tinau's stage one was completed and 50 kW of power began flowing to BTI. A visiting foreign engineer commented that everything was "very haphazardly connected together, but it worked!" At last the diesel generator fell silent and BTI staff could sleep under fans at night.

THE BUTWAL TECHNICAL INSTITUTE

Meanwhile, in the four years it had taken to bring Tinau's stage one to completion, the Butwal Technical Institute (BTI) had also gained its footing. In the spring of 1965 BTI began with just four trainees and ended the year with fifteen. That number rose to twenty-five with new admissions at the beginning of 1966. Initially there was no fixed program with trainees working eight hours a day in various workshops on tasks as needed. This work paid for their living expenses—simple food and a bed in a hostel attached to one of the missionary residences. Six days a week a trainee's day began with an hour of classroom study, from 6:00 to 7:00 a.m., followed by four hours of labor in the workshops (7:00–11:00 a.m.) with another hour of classroom work until noon. Classroom topics included mathematics, basic design engineering, English, and "general knowledge." After an hour of lunch break and four more hours of work, the workshops closed down by 5:00 p.m. Then it was back to the hostel where boys took turns cooking for others in the group. The results were often intensely disappointing from a culinary standpoint and also challenging when upper-caste trainees were asked to eat food prepared by lower-caste colleagues. Lights out happened at 8:00 p.m. In his annual report for 1965 (Møgedal 1983: 12), Hoftun noted, "We have very little trouble with the boys. When they are not working they are so tired that they sleep." Quite a few trainees quit, but "[t]hose boys who have not dropped out are hard-working, interested, well behaved, and really an inspiration to all of us."

Hoftun was unapologetically tough on the young BTI trainees, usually fourteen–eighteen years old when they enrolled. By the end of 1966 more than four out of five students who had started the program had dropped out. "They thought our training was brutal," said Hoftun.

> It *was* tough, it was hard on those boys. And maybe it was a bit *too* tough. But those who were trained that way, and stuck with it, they were good people. It was a way of selecting the best. That was the plan from the very beginning.

When I asked Hoftun how they selected trainees for admission and whether they gave preference to low castes or the poorest of the poor, he stressed that their admissions criteria were strictly practical. "We just made rules that all were equal. We never asked about caste. They were tested for gifts in things like dexterity, resourcefulness, confidence, concentration, stamina, and so on." As for school learning and having passed the school leaving certificate (SLC),[3] "We didn't ask for that. We rather considered that to be a *dis*-qualification!" Hoftun noted that some criteria, like fitness and hand skills, tended to disadvantage high castes while low castes were sometimes cripplingly timid. "We had both Nepalis and foreigners on the [admissions] committee and they jointly judged the character of the person, whether he would be one who would be expected to do well in this tough life." In general he found applicants from the hills to be more rugged, self-sufficient, and resourceful than those from the Tarai (with the exception of Tharus who were "honest, simple, and worked hard"). Hoftun, who grew up in the mountainous interior of Norway, noted that this upland–lowland divide "isn't only in Nepal. In Norway it's some of the same thing."[4]

---

[3] "School leaving certificate"—acquired after passing a general education examination at the end of ten years of study.

[4] If nothing else, Hoftun's hill versus plains assumptions prove that this kind of politically loaded stereotyping is not unique to Nepal. The difference is that, whereas in

Many of the young Nepali trainees who came for the exam were very unclear on exactly what BTI was about, a problem that speaking with foreigners sometimes only made worse. Rudra Bahadur Chhetri (who later took on important administrative positions in BTI and other BPC-related companies) remembers when, in 1965, he and a group of friends showed up at Odd Hoftun's bus body office looking for information on the institute and how to join. In broken Nepali Hoftun told them a date and something about a *jahaj*, the Nepali word for ship or airplane. Thinking that they were about to become pilots the boys returned on the given date only to find a *janch*, or entrance exam, awaiting them. Speaking Nepali with a Norwegian accent soon became a favorite pastime for trainees (Møgedal 1983: 16)! Another former trainee told me that when he showed up in 1965 he had been told about "technical training" but honestly had no idea what it entailed.

Once enrolled, trainees looked at Hoftun with a mixture of fear, respect, admiration, and affection. One former trainee remembered referring to Hoftun as Jung Bahadur: "Jung Bahadur ayo!" signaled a sighting of the stern, serious tyrant.[5] Even if Hoftun was not entirely to blame for the austerity of trainee life, he did not seem to oppose it and, as director of BTI, came to be seen as its author. Few had the nerve to look him in the eye. But no one could accuse Hoftun of living in luxury at their expense. Working seemingly endless hours in his bus chassis office, eating simple food, and cycling everywhere, one person jokingly likened Hoftun to a *brahmacharya*—a South Asian religious renunciant who shunned earthly pleasures. Over time most came to respect Hoftun for his self-discipline, fairness, and unbending commitment to ethical practice. A few who got to know Hoftun personally saw him as a guru or even a father figure. Hoftun held himself to high standards and expected others to do the same.

Although UMN's agreement with the government of Nepal officially forbade its staff from engaging in proselytization, some BTI expats wanted to make Christian education available to the trainees. They sponsored evening

---

most parts of the world the lowlanders have power to disparage and disadvantage the "hillbillies," in Nepal the equation is reversed.

[5] Jung Bahadur Rana was the ruthless but pathbreaking founder of the Rana dynasty that ruled Nepal from 1846 to 1951. Thanks to Tara Lal Shrestha for this anecdote.

and Saturday Bible classes but made a point of at least trying to separate these events from official BTI spaces and activities. Trainees should feel welcomed, but not pressured, to participate. Decades later I spoke with an independent Nepali researcher who interviewed many early BTI trainees and was genuinely surprised when none of them reported having felt coerced to join Christian activities. A few trainees did become Christians (having also been influenced by Indian Christian missionaries active in the Butwal area) but many more came to a deeper understanding of Christianity even while retaining their (usually Hindu) religious identifications.

Official BTI training may not have been overtly religious but it was very intentionally ethical. Once a week during the "general knowledge" part of their classroom study Hoftun laid out the basics of a positive secular business ethic. Hoftun saw ethical behavior as the only way to survive and thrive in a market-based industrial society—the only future he foresaw for Nepal. "Gradually I had to come to grips with how to convert people without converting them," Hoftun told me. Values that for him had religious valences could also be presented in secular, rational terms. Hoftun explained,

> For example, "truth," [I would tell trainees that] if you lie about a product or an action, that will not last for long. The product will fail or the truth come out [and you will pay the price]. So truth is the basis for all development. It's all very logical. Lying is not a way of doing business. It won't last for long.

Ultimately, Hoftun argued, dishonesty was not in anyone's rational self-interest and, at a societal level, it only held a country back. Truth and its derivatives (transparency, dependability, accountability, accuracy, hard work) were the only solid foundations for personal, professional, and national development. Reflecting on his experience of Hoftun's moral education, one former trainee probably spoke for many when he said,

> What Odd Hoftun taught us was some kind of universal teaching that was needed for development, to become a mature person, to be understanding and be equipped for life. Topics like "motivation" he used to teach us. What does it mean? How can we be motivated? What makes us motivated?

Topics like "honesty." How can we demonstrate honesty in life, in our work? And like that. I must say they were very important elements for young people like us in those days. We were going to become something in the future and we wanted to achieve something in our lives and those elements were very useful.

For Hoftun solid ethical values were as important as solid technical skills for the young tradesmen he was preparing to enter, and in many ways *establish*, the capitalist marketplace in Nepal.

GROWTH AND COMPLEXITY

The Butwal Technical Institute (BTI) grew both in physical size and output. Volunteers in Europe continued to ship tons of equipment and supplies. Trainee-made products (furniture, roof trusses, metal tanks, and so forth) were making profits. BTI served as the main basic equipment supplier for the ongoing Tinau hydropower project and, after 1970, was also the Butwal Power Company's (BPC's) main customer. By the end of 1967 trainee numbers had climbed to forty-eight and the institute had over 1,500 square meters of workshop and office area, and 1,200 square meters of living quarters.

By the late 1960s, BTI hired its first Nepali professional instructors. R. P. Sharma was part of the wave of Nepalis fleeing the nationalist military takeover in Burma—where Sharma had become a highly skilled master welder. "He could keep the quality that could be depended on and trained his people very well," Hoftun recalled. Thanks to Sharma BTI could produce high-quality power plant components like penstock pipe and valves. D. R. Rana was working as a master mechanic and trade instructor in India when he met Hoftun who immediately offered him a job and then practically shamed him into joining BTI out of duty to his country! Both Sharma and Rana were pillars of the BTI training program until their retirement decades later (Svalheim 2015: 84).

As staffing levels grew and conditions normalized within BTI, emphasis gradually shifted from crisis management to debates over vision and proper practices. Hoftun—as BTI's founder, visionary, and director—was never far from the center of controversy. There seemed to be three basic areas of disagreement. First, and probably most important, was the debate over resources and standards. What level of professional standards should BTI strive to achieve and at what price? Several of the expat staff were highly skilled German professionals whose careers before coming to Nepal had been committed to training and production regimes that tolerated nothing but perfection. But high quality came with a high price tag. Should BTI import the latest precision machinery capable of the highest quality output? Many felt so and argued that settling for less was an insult to Nepal and Nepalis. But others, most notably Hoftun himself, felt that it was much more important to have simpler (often used and somewhat outdated) technology that could be maintained by BTI itself at a relatively low cost, even if the resulting output was not up to First World standards. Hoftun's commitment to (to others, obsession with) sustainability within Nepali capacity trumped the urge to leapfrog ahead. He was also committed to cultivating the creative spirit of "making do" that was necessary when dealing with less-than-ideal circumstances. Nothing kills resourcefulness, creativity, and sustainability like too much money, Hoftun argued.

Hoftun's embrace of austerity also had implications for quality of life for both staff and trainees. Many expats criticized the difficult living conditions that trainees were expected to endure and were often unhappy about their own living conditions as well. Life in Butwal—with its sense of isolation and extreme heat for much of the year—took a toll on expats, some of whom should probably have never come in the first place. Hoftun argued that austerity was simply the result of short funds but others sensed (correctly) that Hoftun was trying to impose his own Spartan discipline onto them while not trying hard enough to raise additional funds. Even the trainees eventually went out on strike demanding better living and work conditions.

A second point of contention between Hoftun and many of the other expat staff concerned curriculum and teaching philosophy. Through a combination of necessity and conviction Hoftun favored a more unstructured, generalist approach to training that would allow graduates to adapt to changing market demands. "It is important to remain flexible in our approach and not put too much of our efforts and resources into any single line of activity," Hoftun argued (Møgedal 1983: 13). But others felt that standardized curriculum and rigorous specialization was the only way to produce really highly skilled tradesmen. A related debate concerned the relative emphasis to be placed on either "training" or "production." Hoftun favored a balance of skills acquisition and production for the market while many foreign staff favored one or the other.

All of these debates ultimately pointed to a third controversy: Hoftun's leadership. Hoftun was both director of BTI and general manager of BPC which was deep into the Tinau project (for which Hoftun was also chief design engineer). In addition to trying to juggle essentially two, maybe three, full-time jobs, Hoftun's perfectionist tendencies meant that he was inclined to micromanage those below him and not good at delegating responsibility. His belief in relatively nonhierarchical administrative structures meant that, as the institutions grew, Hoftun ended up trying to supervise far too many people. Always stretched to the limit, many experienced Hoftun as rigid, blunt, and quick to anger. By 1970 the expat staff in Butwal was in open rebellion and Hoftun submitted his resignation from the BTI directorship to the United Mission to Nepal (UMN) board (Møgedal 1983: 23). "That was the toughest time I had in Nepal," Hoftun commented. When I asked if he had any regrets, he said simply, "Leadership style. There were many things to regret [about that]. There was too much, too many people reported directly to me. It was beyond the capacity of one person." But it was more than just being overextended. Hoftun simply lacked the "people skills" (listening, negotiation, flexibility) to hold the fractious institution together.

In retrospect, the trauma leading to Hoftun's resignation from the BTI directorship seems to have triggered a significant reorganization and expansion of the institutions that he had founded. During the late 1960s and early 1970s BTI and UMN administrators refined the institute's management structures writing up a new constitution, establishing management committees, and formalizing a Management Board under UMN supervision. Many of these changes had a direct impact on the BPC story. UMN renewed its efforts to recruit Nepali board members and drew up a timetable for the eventual handover of BTI's management and ownership into Nepali hands. After resigning from BTI, Hoftun could focus all of his energies on his remaining post as general manager of BPC while also working on new initiatives. As for BTI, UMN hired two successive expat directors, both of whom cracked under the strain of trying to recreate European institutions in Nepal.

Eventually UMN hired Nepali directors for BTI: first, Soviet Union–trained engineer Dinesh Upadhyay and then, after Upadhyay's untimely death, Simon Pandey, himself a former BTI trainee and instructor. As more and more Nepalis replaced expats in teaching and administrative posts, institutional tensions diminished as BTI more effectively adapted to local conditions even while retaining its original mission and values. For the first time UMN expats found themselves reporting to Nepali superiors, a condition that took some getting used to on both sides. Foreigners had to get over a sense of privilege and get used to slightly more formal and hierarchical leadership styles, and Nepalis had to get used to foreigners accustomed to expressing their opinions and suggestions directly. As ever, language proved to be a challenge as signals could sometimes be misconstrued when both sides were using a language (English) that was not their mother tongue.

After completing stage one of the Tinau project in late 1970, BPC took on the much more challenging stage two which involved digging an additional 1,200 meters of tunnel, excavating an underground powerhouse cavern, and installing sequentially three larger turbine-generator sets. During stage one BPC was responsible for all of the design, purchasing, accounting, administration, and construction coordination that went into the Tinau project—with Hoftun shouldering much of this work himself. This situation had to change and therefore it is at this point that we begin to see the first institutional divisions of labor that would—in the coming decades—gradually result in the separation of BPC from its specialized corporate progeny: Hydroconsult, Himal Hydro, and Nepal Hydro and Electric (NHE). Hoftun explained to me that, in Norway, the hydropower industry was divided into four main sectors: (1) power companies that were involved in developing, managing, and distributing electricity; (2) engineering design services that drew up the technical plans for new projects; (3) civil contractors who built the physical infrastructure (from dams to tunnels); and (4) electromechanical suppliers who produced the complicated pipes, valves, turbines, generators, control panels, and distribution systems necessary to actually produce and sell power. According to this logic, BPC should have been in sector 1 but instead, by necessity, also encompassed sectors 2 and 3, and (indirectly) as much of sector 4 as BTI could manage to produce (with the rest of the hardware arriving used from Europe). Because each of these sectors has its own unique characteristics and relations to the broader market, it is logical and desirable to have their functions institutionally separated.

The first instance of this separation was the creation of the Development and Consulting Services (DCS) division in 1972. The original agreement that UMN signed with the government in 1963 had envisioned an "Institute of Technology and Industrial Development" with two divisions: a training facility that became BTI and a design and development service. The latter sat dormant until 1972 when Hoftun launched the new division, with himself as its first director. Although the DCS's mission was much broader than just working with BPC,[6] under Hoftun one of its immediate jobs was to take over key elements of the Tinau project. Specifically, the DCS contracted with BPC to do all the technical design work for Tinau, and they took on the job of hiring and supervising the project's construction workers. By separating out the design and construction functions (that would later be further institutionalized into Hydroconsult and Himal Hydro), BPC could focus on being a power company: distributing and selling electricity to consumers and looking ahead to developing new projects.

Of course BPC's primary customer was the technical institute itself. Even before the first stage of Tinau came on line in 1970, BPC was selling power to BTI from its diesel generator. But BPC had also already started marketing power to fifty other local consumers including the government hospital, government offices, and three industrial users. Therefore, by the time the first hydropower came from the 50 kW Tinau generators, BPC already had a tiny consumer market and demand quickly grew to meet supply. Across Butwal electric lights began appearing in private homes, streetlights lit up the night, and small businesses popped up based on newly available power.

---

[6] In the coming years, DCS also did important work in designing biogas digesters, micro-hydro components, and many other appropriate technology projects.

Dealing with consumers, or at least dealing with them well, required human relations skills that were not Odd Hoftun's strong suit. Expanding sales for BPC meant installing distribution infrastructure—power poles and lines—throughout the Butwal bazaar. In one still-remembered incident, crews digging holes for poles encountered an irate landowner who refused to have power lines over his property.[7] When the poles arrived for installation, the angry man ordered his long-suffering wife to stand in one of the post holes. Faced with this dilemma, Hoftun, in rather undiplomatic fashion, simply had the woman removed, the pole erected, and moved on. After this the landowner renamed his dog "Odd Hoftun" and enjoyed calling for his furry friend every time the human Hoftun was within earshot![8]

Again stretched to the limit by an expanding workload, and often forced outside of his administrative comfort or competence zone, Hoftun was delighted when a young Nepali engineer arrived at his office in September 1972.

> I remember well the day in Butwal when Balaram [Pradhan] first walked into my empty office in BPC asking whether I had a job for him. For me it was like a gift from heaven, and I appointed him immediately as my personal assistant. Within a year he took over as [BPC's] General Manager with me serving as advisor.[9]

So began a close personal relationship between the two men that lasted almost forty-five years. That day Pradhan was twenty-four years old and had recently finished a BSc degree in electrical engineering from the Institute of Technology, Banaras Hindu University, in India. Hoftun was forty-five and although the two were in some ways very different, their differences were complementary: whereas Hoftun was introverted, shunned publicity, and preferred to let his actions (hopefully) speak for him, Pradhan was warm,

---

[7] Again, at the time there were no legal precedents or regulations in place to determine how things should be done.

[8] Thanks to Tara Lal Shrestha for this anecdote. A version of this story also appears in Svalheim (2015: 159).

[9] From an email from Odd Hoftun to Dhiraj Pradhan, February 2, 2016.

gregarious, patient, a gifted communicator, and had an incredible knack for remembering people and maintaining relationships.[10] Where the two shared common ground was in their ethical impulses. Pradhan was an idealistic, left-leaning progressive with a strong commitment to fairness and national service. In Hoftun he found a kindred spirit and something of a guru or father figure—with all the closeness and sometimes frustration that such a relationship entails. Pradhan and Hoftun would work together on almost every major BPC project from Tinau through the grueling privatization process that ended in 2003.

But most immediately, Balaram Pradhan was the person needed to develop the "public" side of the BPC's public utility mission. He quickly became the public face of BPC, interacting with everyone from homeowners and shopkeepers to industrialists and government officials. He was BPC's chief salesman but also an ambassador working to establish a good name and reputation for the company. Pradhan walked door to door through Butwal soliciting customers, answering questions, and solving problems. He was soon on good terms with virtually everyone and even decades later many in Butwal fondly remember Balaram Pradhan as the person who helped them secure power for their fledgling businesses. By April 1974 BPC had 604 domestic and 22 industrial customers, and powered 121 streetlights at night. But with only 50 kW of hydropower (and the diesel generator for peak periods) to distribute, demand quickly outpaced supply. Pradhan worked out deals with industrial consumers that they had to stop drawing power at 5:00 p.m. to make electricity available to shops and homeowners during the period of evening demand. On many occasions Pradhan had to literally walk around town making sure that factories and workshops shut down on time.

Pradhan had excellent "people skills," including the ability to get tough when necessary. With power going to government offices and streetlights, the Butwal municipality was one of BPC's main customers. But Butwal's mayor

---

[10] When I first met Balaram Pradhan in June 2015, reading from my business card he immediately pronounced my name correctly (something that virtually no one does, anywhere in the world) and asked if I knew Eric Liechty, a man he had known briefly forty years earlier. Eric is my first cousin and had worked as a "Pax" volunteer at BTI.

harbored deep suspicions of BPC's motives. Seeing the foreigners involved in it as part of some neocolonial plot to exploit Nepalis, in 1973 the mayor refused to pay the city's electric bill. When negotiations failed, Pradhan personally cut the city's power supply. In retaliation the mayor cut the water supply to the BPC office and a standoff ensued, lasting for several months. But Pradhan managed to keep the moral (and public relations) high ground such that the mayor eventually backed down and subsequently lost the next election (Svalheim 2015: 160). The point is that Balaram Pradhan had the ability to lead BPC through the cultural and political shoals it encountered and helped to establish its reputation for service and reliability. He also made BPC a commercially viable business.

TRAINING VERSUS EDUCATION

From the very first time that Odd Hoftun approached the government of Nepal with the idea of establishing a technical institute there was debate over whether such an institution would be involved in "education" or "training." Initially the government itself resisted the idea of a "school" (fearing missionary indoctrination) and only agreed to the proposal when it was pitched in terms of industrial development (under the Ministry of Industry). But in 1972, with the Butwal Technical Institute (BTI) now well established and successfully producing skilled tradesmen, government officials approached BTI administrators with the proposal that it be brought under the Ministry of Education and reorganized as a university. This suggestion touched a nerve that can stir Odd Hoftun's passions to this day. Then and now Hoftun drew stark distinctions between education and training, students and apprentices, theoretical and practical. "We *hated* the term 'school'," Hoftun told me.

> That was a word we *never* used. And when people came from outside talking about a "school" we were right away telling them that this is *not* a school! It is *industrial development*. It is economic development in the widest meaning.

As an academic, I was rather taken aback by his visceral distaste for education and asked him to explain. Interestingly, he began his answer by referring back to the Norway of his youth, again describing its extreme backwardness as compared with the rest of Western Europe. He also again noted Norway's historical experience of being exploited (and, in fact, colonized) by its neighbors. The implication was that, while education might be fine and good, what Nepal lacks—as Norway had lacked—was people with marketable skills. Skilled people will enable industrialization which will enable Nepal, like Norway, to rise to a position of national self-respect and independence.

For Hoftun a school is a place where everything is all set up and ready to go, everything is provided: a curriculum, teachers, a building. "It is isolated and closed in." In education there is

> nothing systematic, nothing going into the marrow. We have seen so many examples of this in Nepal. They are educating people who are not employable.... [Schools are] usually theoretical and not training people for practical jobs under simple and difficult circumstances.

By contrast training is about being integrated into the community, about learning how the real world works. "They need training in practical skills to do what needs to be done." From the beginning BTI training stressed the need to work with, and be responsive to, an ever-changing market.

Hoftun also saw the trainee or apprenticeship approach to be empowering in ways that schools rarely were. By learning to transform hard-earned personal skills and labor into objects and services with marketable value, people could break away from the cycles of defeatism and dependence that were often ingrained in Nepal's agrarian, caste-based society. As in Norway, where agriculture had been almost the only employment option for many people but where there was too little land for too many people, so also in Nepal Hoftun saw industrial training and industrial jobs as a whole new employment sector that would allow Nepalis to stay at home, earn living wages, and build the economy. Industrial training gave people a way out of the trap in which land was the only real productive asset, but land was in the hands of only a few. A skilled tradesman became his own productive asset.

BTI's goal was to work with the community, and eventually local industry (almost all of it started by BTI tradesmen themselves), to prepare people for jobs that existed. Hoftun explained,

> The secret of success of BTI is that it was developed together with industry and has been operating very close to the industry providing apprenticeships for the trainees. When there is no connection to an industry, vocational training becomes a school. BTI was never a school. (Hoftun 2011)

BTI aimed to produce tradesmen equipped with basic business skills needed to operate independently in the market. The institute graduated dozens, and eventually hundreds (and now thousands) of tradesmen. Many of them started as employees before launching their own workshops (often training their own apprentices). Many became very successful and some became rich—a fact that drew criticism from some of Hoftun's fellow missionaries. "We were blamed for that by our colleagues in UMN," Hoftun remembered. "Someone said, 'These people you are training there in Butwal, they are becoming *rich people*! So you are completely in the wrong line.'" They blamed you for taking people out of poverty? I asked. "Yeah, that was said several times." But Hoftun admitted that he too had some misgivings about unleashing an industrial juggernaut in Butwal with all the attendant problems of class disparity and labor exploitation that go along with capitalist business. But he insisted that BTI produced tradesmen, not businessmen. "Well," he sighed, "it's a dilemma. It's no easy thing."

Hoftun and many other staffers were dead set against turning BTI into a school and when they made their case to the government, they pointed to a number of other practical obstacles to such a change of course. First they pointed out that, unlike a school which either students or someone else had to pay for, BTI was self-supporting, which allowed it to admit trainees at little or no cost. In the course of their training, trainees made things and provided services that were of marketable value. From producing furniture to metal hardware to bridge components to welding-repair services, trainees paid for their training by learning to deal with customers in the market. In effect BTI was an industry itself, something a school could never be. Government officials suggested that the institute could go onto a semester system, but when BTI administrators "asked what our customers should do with their needs in between semesters, differences from the normal academic training patterns were realized" (Møgedal 1983: 30).

By the mid-1970s Hoftun and others in the BTI/BPC community had learned a useful lesson (one that Nepalis had probably known for a long time): when confronted with government officials who wanted you to do something you did not want to do, stall for time and mobilize your networks. "We who had worked in Nepal," explained Hoftun, "we knew that things could be changed. By having good contacts and talking with people [in positions of authority], and by just holding out and waiting, we could negotiate with them a better deal." Today's annoying official or official policy might not be around tomorrow when things might be more amenable. If not, you could hold out some more.

The government's suggestion that BTI be transformed into a university under the Ministry of Education was one such annoying policy that had to be resisted. BTI held off the unwanted advances for a while but as 1978 approached—the year in which BTI was to be handed over to government control as stipulated in the original agreement—questions of how BTI would be managed became paramount. An independent consulting group (New ERA) hired to evaluate BTI's performance found that a high percentage of trainees found good jobs in their areas of specialization and that the school's "apprentice-style, production-oriented" approach was effective (Møgedal 1983: 37). Sales of BTI-produced goods were routinely earning yearly profits capable of covering the institute's operating expenses. The consultants' report heralded BTI as a model organization. While the BTI handover should have happened in 1978, negotiations with the government (over things like board membership and whether BTI should fall under the Ministry of Education or the Department of Cottage Industries) dragged on for years. Finally in 1982 the government just requested the United Mission to Nepal (UMN) to continue running BTI until details of the transfer could be worked out.

Hoftun and other BTI and UMN staff (both expats and Nepalis) may have resisted handing the institute over to government hands, but it was not because they wanted to keep BTI under foreign control. In fact BTI had been under Nepali directorship since 1975 and most of the other administrative and instructional staff were Nepalis in the early 1980s. UMN continued to place foreign volunteers in relatively short-term advisory and technical roles (in BTI but especially in the Development and Consulting Services [DCS]). What was at stake was how to negotiate the transfer in ways that minimized changes to the institute's pedagogical approach and philosophy. Ultimately it was under Nepali leadership that BTI was able to negotiate an agreement with the government whereby BTI became officially government owned, but its governing board remained independent. Today BTI's board includes representatives from various relevant stakeholder groups that have resolutely maintained the institute's industrial apprenticeship model. BTI became coeducational in 1992. With a smile of contentment, Odd Hoftun told me, "After all these years BTI is still in private hands."

WRAPPING UP TINAU AND THE FATE OF BPC

Whereas it had taken four years to dig 400 meters for the Tinau project's stage one, it took only three and a half years to dig the next 1,200 meters of tunnel, along with the underground powerhouse, for stage two. With the Development and Consulting Services (DCS) taking over the project's design and construction responsibilities, and with more experience, more labor, and better equipment (especially pneumatic drills), the Butwal Power Company (BPC) was able to install turbines (used but refurbished by the Butwal Technical Institute [BTI]) producing 175 kW in July 1974 and a second matching unit a year later, bringing its total production capacity up to 400 kW. By July 1976 BPC had 824 domestic and 47 industrial customers in Butwal, and powered 261 streetlights at night.

Stage three involved an additional 800 meters of tunneling and building a small dam. By significantly increasing the amount of available head (the water's vertical drop and therefore generating power), and by installing a third large turbine in the powerhouse, these changes increased the project's generating capacity to just over 1,000 kW when the project was completed in May 1978. King Birendra officially inaugurated the project in February 1979.

By then BPC powered over 1,000 domestic and 65 industrial customers, as well as more than 300 streetlights.

With Tinau finally completed, the next question was how to handle the power plant's transfer of ownership to the government. This had been the plan from the beginning and was part of the original agreement with the government that had led to BPC's incorporation. Initially Hoftun had been operating on the common assumption that power production and distribution should be government affairs, as he was familiar with in Norway and as they had been in Nepal up to that point. By 1978 he was having second thoughts about turning over the product of so much hard work for free, but it had become inevitable. In July 1980 the Nepal government took over the Tinau power plant, the entire distribution network that BPC had developed, and an administrative staff to keep it running.

But what would become of BPC? If BPC was created in order to build Tinau and launch its distribution system, was its mission now over? The government thought so and sought its liquidation. But Hoftun, Pradhan, and others had other ideas. Already before the Tinau project was completed, Hoftun was doing preliminary work on a new, larger hydropower project between Tansen and Pokhara near the confluence of the Andhi Khola (*khola* means river) and the Kaligandaki River. But that Andhi Khola project was still more of a dream than a plan—not something that BPC could bank on, much less use to justify maintaining staff. Still, BPC leaders wanted to maintain the company's ties with BTI. Having provided increasingly specialized hydropower construction equipment for Tinau, it would be a pity to see those skills (and that market) abandoned. Likewise with the large number of now experienced construction workers. Why not keep them and others working in the hydropower sector?

In the negotiations leading up to the Tinau project's nationalization, Balaram Pradhan took the lead in trying to preserve BPC's institutional life. "He had to fight *hard* for keeping the company as a paper company for the time being," recalled Hoftun. "We had the next project in view but not firmly planned. So in those final negotiations [Pradhan] was handling all that." Fortunately Pradhan succeeded, though for a time BPC existed only as a shell, a "paper company," with NPR 50,000 in the bank and a board of directors but no employees or other assets whatsoever.

During the Tinau project's final years and months, while Pradhan was fighting to keep BPC alive, Hoftun and others were trying to do the same for other aspects of an embryonic hydropower sector. Hoftun recalled how,

> when the question of handing over came up, we felt that here we had now a crew of workers and technicians who should continue to have a job and expand further. So then the first thing we started on was setting up Himal Hydro as a contracting construction company, taking over the bulk of the people who had been employed on the construction work. So they were transferred then from being employees of BPC to become employees of Himal Hydro, established as a separate organization. They would have to go out and look for jobs immediately afterwards in order to keep their people employed.

In February 1978 the United Mission to Nepal (UMN) formally registered Himal Hydro and General Construction Company as an independent private entity. At a time when BPC's own future was still in doubt, spinning off Himal Hydro would at least make sure that its construction expertise was preserved. Then, and for decades to come, Himal Hydro was the only company in Nepal with tunneling capabilities though it also competed with other companies for building and road construction contracts, installing power lines, and so on.

Himal Hydro's origins were, like those of BPC a decade earlier, virtually unprecedented. Two-thirds of Himal Hydro's initial investment capital came in the form of an outright grant from Norad, the Norwegian aid agency: the Norwegians provided 2 million Norwegian krone (NOK) (c. USD 300,000), with the Nepal government investing the equivalent of NOK 1 million. And whereas the Nepal government claimed one-third of Himal Hydro's investment shares, the Norad funds were a simple grant to UMN which became the majority shareholder with two-thirds. Hoftun explained,

> To finance a company by grant funding, that was something very extraordinary for Norad. Such a *risky* type of business as contracting for a

[hydropower] civil works contract, you can go bankrupt in no time if you have misjudged. So it was a maximum risk in setting up a company for this purpose.

In other words, investing in any company whose business depends on contracting for tunneling work is extremely risky. Tunneling companies have to estimate their expenses up front based on tests and best guesses. But ultimately the rock conditions inside a mountain will never be known until the project is well underway. If conditions are as anticipated, one can expect within-budget costs and a profit. But if they are worse than expected, time and cost overruns can quickly spell bankruptcy.[11] When Hoftun added that such a grant or investment "was against all principles," I asked how they had convinced Norad to do it. "Well, we had friends," he explained. "One was a top-level engineer [in Norway]. He, as a Christian, was very much interested in mission work, so he wanted to help us. So we asked him to take up the job to persuade Norad to give this money." Convincing the Nepal government to make an unprecedented investment of this sort fell to Balaram Pradhan as part of the larger negotiations he undertook to nationalize the Tinau project. "That for them also was quite a big deal, quite a big decision," explained Hoftun. "But in the end [the government] saw the potential of doing this."

During the years that Tinau was under construction, the workshops at BTI had been producing a variety of project equipment from wheelbarrows and tunnel trusses to penstock and hydraulic valves. For practical accounting reasons, in 1977 BTI segregated this metal and mechanical fabrication component of its operations unit into a separate, free-standing company (as they had BTI's wood production unit) named Butwal Engineering Works (BEW).[12] BEW took over the functions of BTI's machine and welding shop that had been busy fabricating specialized products for the Tinau hydroelectric project. But like Himal Hydro, BEW was involved in much more than just products for the hydropower sector. With the East–West Highway then

---

[11] As Himal Hydro discovered at Khimti. See Chapter 5.
[12] BTI held two-thirds of the new company's shares while UMN held the remaining third (Møgedal 1983: 38).

under construction, as well as the road linking Butwal and Pokhara, BEW's services were in great demand for repairing heavy equipment and building bridge components. They built large towers and lightweight tubular poles for electrical transmission. They even made industrial parts for a new "scotch whiskey" distillery in Bhairawa. BEW soon became an important employer of BTI tradesmen and even began paying BTI to train people for future employment with BEW.

Toward the end of the Tinau construction period Balaram Pradhan had taken a year off from the BPC directorship to do a master's degree in electrical distribution from the Norwegian National Institute of Technology in Trondheim.[13] After his return and his work negotiating with the government to preserve BPC, Pradhan had a choice. With BPC essentially in mothballs, he could stay with the Tinau power plant and become a Nepal Electricity Authority[14] (NEA) employee. He could stay somewhere within the constellation of organizations related to BPC (UMN, BTI, DCS, Himal Hydro, BEW) and continue working closely with Hoftun. Or he could go into business on his own. For Pradhan government employment was out of the question because he feared it would force him to compromise his ethical principles. Initially he chose to work with BEW, first as a development engineer and then, from 1981 to 1983, as its executive director. But, according to Pradhan's family members, staying within the UMN family of enterprises was not entirely appealing either. Just as a number of other Nepali professionals noted to me, Pradhan felt that institutional "missionary values" kept salary and perks to a level where his best interests were not being served. With a young and growing family to support, in January 1984 Pradhan launched

---

[13] Pradhan was the first of now hundreds of Nepali engineers to receive advanced training in Norway with Norwegian government assistance.

[14] Technically in the early 1980s the government entity to which BPC handed over the Tinau project was the Nepal Electricity Corporation, the forerunner of the Nepal Electricity Authority (NEA). NEA came into being in August 1985 as stipulated by the Nepal Electricity Authority Act of 1984. For more details, see Shrestha (2016). To reduce confusion, I will refer to the government's electricity authority as NEA even before that entity's official birth in 1985.

his own private engineering services company in Butwal, despite Hoftun's efforts to retain him.

CONCLUSION

Once the Nepal government took over the Tinau power plant in 1980 and its staff became Nepal Electricity Authority (NEA) employees, everyone involved in the project's implementation held their breaths to see how operations would proceed. As of 1983, when Tor Møgedal wrote his short history of the Butwal Technical Institute (BTI), things were progressing smoothly and even after major challenges (for example, a huge flood that damaged the facility in 1981) the Tinau crew was able to return to production in good time. NEA independently contracted with Himal Hydro for construction assistance and things proceeded in a businesslike manner (Møgedal 1983: 48). But as years and decades passed, Hoftun began to lose his basic belief in a national government's rightful role as producer and distributor of electrical power. He and others watched as basic operations at Tinau began to deteriorate. Not only was NEA unable or unwilling to upgrade Tinau (by, for example, installing more efficient low-volume turbines for dry season production), they sucked so much profit out of the enterprise that they no longer even budgeted money for routine maintenance. Even though high-quality and affordable maintenance and refurbishing services were available nearby in Butwal, NEA adopted a wait-till-it-breaks approach—by which time damage was more severe and repairs more costly. All or parts of the plant sat idle while waiting for the government to allocate repair money, wasting even more money and causing more local shortages. Amazingly, when Butwal businesspeople approached NEA offering to buy or lease the Tinau power plant (in order to manage it properly), NEA refused (Hoftun 2011). Watching the Tinau plant decay under government supervision helped turn Hoftun's attitude increasingly in favor of private sector ownership for power facilities.

But if Tinau's fortunes declined, the opposite is true of Butwal town as a whole (see Fig. 2.1). Already by the early 1980s Butwal was a very different place than it had been when BTI began in 1964. From a small town built around entrepôt trade between hills and plains with a population that

Figure 2.1 Odd Hoftun Street, Butwal
Source: Photo by Naomi Liechty.

fluctuated seasonally, by 1983 Butwal was becoming a center of industrial production. BTI and its daughter companies alone employed some 500 people and the surrounding area was full of other manufacturing enterprises started and staffed by BTI-graduated tradesmen (Møgedal 1983: 49). Since then Butwal has become the major exception to the otherwise gloomy picture of industrialization in Nepal. It is a major industrial hub and, arguably, the capital of electricity-driven industrialization in Nepal. Thousands of BTI tradesmen have started hundreds of workshops and factories that, in turn, hire still more BTI graduates. Once on their own in the market many BTI tradesmen competed with BTI itself, forcing elements of the Institute's industrial production "out of business."[15] Some have started trainee apprenticeship programs of their own and others are making contributions to civic life in Butwal, such as sponsoring education programs for the poor.

One former Butwal Power Company (BPC) engineer I interviewed summed up what he thought Odd Hoftun's main contribution to Nepal had been:

> Hoftun saw that there was no other way but to train local people to do skilled work. They started training people and that's how they established all those industries [in Butwal]. Their point was not to earn money but to train people. They cooperated with the government to industrialize the nation. All over Nepal now we have so many trained persons, technicians who have graduated from BTI.

Hoftun believed that if industrialization had been a key element of building dynamic western economies and raising standards of living, then the same would have to happen in Nepal. Though his positions were often deeply suspect to other missionaries and other religiously motivated aid workers (who felt more comfortable contributing to things like health care, education, or rural development), BTI's philosophy revolved around the idea that Nepali engagement with capitalist market forces was inevitable and that equipping workers to succeed in that market was a way of fighting poverty and creating individual self-sufficiency. Even though many saw the market and industry as dehumanizing and even enslaving, Hoftun believed that, if and when pursued ethically, the market could be a means of social advancement.

---

[15] For example, BTI bridge-building and micro-hydroturbine manufacturing (designed by DCS and built by BTI trainees) could eventually no longer compete with BTI graduates independently producing the same or similar products at lower cost.

This chapter has chronicled BPC's relatively unplanned embryonic stages of development, along with those of Himal Hydro and what would later become Hydroconsult and Nepal Hydro and Electric (NHE). Obviously a great deal of commitment, creativity, and determination went into these early stages (and those yet to come) but it is important to clearly acknowledge that this is more than an "up-from-the-bootstraps" story of foreigners and Nepalis working with unbending resolve. Building new companies, new hydropower plants, and new human capacity takes hard work but it also takes *money*—money that Nepal did not have and that had to come from elsewhere. Here that money (or value) came in several forms: (1) as outright grants from entities such as UMN, Norad, and various missionary agencies; (2) in the form of deeply subsidized (essentially free) labor from the United Mission to Nepal (UMN) expats who contributed expert skills to BTI, BPC, and the Development and Consulting Services (DCS) at rates of pay vastly below what they could have earned in their home countries; and (3) in the form of outright volunteer labor and donated goods, as in the case of the small army of Norwegian volunteers who gathered, crated, and shipped hundreds of tons of equipment to Nepal.

Odd Hoftun compared the origins of BPC and Nepal's hydropower sector to planting a seed and watching it grow. But he also made clear that the seedling needed water and fertilizer—and perhaps even a good dose of loving care. No doubt Hoftun, Pradhan, and others were careful stewards of the donated cash and labor that came their way, but the fact is that without it—almost certainly—the seed they planted would not have sprouted and grown. BPC's leaders worked to prepare the company for the harsh realities of the capitalist marketplace and its "survival of the fittest" rules. Yet, ironically, they did so from within the decidedly un-market-like protective cocoon of goodwill and benevolent service that thousands of people, in Nepal and elsewhere, worked to build around it. Building BPC required vast amounts of labor, but much of it was in the form of unpaid labor of love.

3
ANDHI KHOLA

Thoughts of a hydropower plant near the tiny village of Galyang on the trail (later road) between Tansen and Pokhara had been percolating in Odd Hoftun's mind for literally decades before planning for the Andhi Khola project began in earnest in the late 1970s. Standing on the ridge at Galyang bazaar in 1959, Hoftun noted how close the Andhi Khola, flowing on one side of the ridge, was to the Kaligandaki River, flowing on the other side. A tributary of the Kaligandaki, the Andhi Khola's riverbed was considerably higher than the Kaligandaki's: a tunnel bringing water through the relatively narrow ridge would have hydropower-generating potential. In 1966 Hoftun asked the United Mission to Nepal (UMN) surveyor who had mapped the Tinau River valley in preparation for construction there to take some basic measurements around Galyang. He found that the two riverbeds were a little less than 2 kilometers apart with an elevation difference of about 250 meters. Some quick calculations suggested that the site had about 5 megawatts (MW) of hydropower-generating potential, about five times the amount available at the Tinau site. Located about halfway between Pokhara and Butwal, a power plant at Galyang would improve the conditions for industrialization in both cities. And, as the Tinau project progressed through the late 1960s and 1970s, Hoftun increasingly saw a new Andhi Khola project as something that would build on and advance the tunneling and construction experience being gained at Tinau (and, in 1978, consolidated in the creation of Himal Hydro) as well as further expanding hydropower engineering and production capacities at the Butwal Technical Institute (BTI), its Development and Consulting Services (DCS), and, eventually, the Butwal Engineering Works (BEW). For Andhi Khola, Hoftun explained, "the intention was to earn a profit and to finance the *next* project or growth of other desirable things like setting up the daughter companies."

In conversations with me, Hoftun made it clear that by the mid-1970s he had started to develop a long-term vision for hydropower development in Nepal. Exactly where and how it would develop was anyone's guess but Hoftun saw how incremental capacity building could create an independent hydropower development sector that would allow Nepal and Nepalis to tap into their country's vast hydropower potential. Waiting for foreign aid to turn that potential into power would only lead to too little, too late, and at the expense of dependence on foreign skills and equipment. Hoftun had started the Tinau project with the relatively modest aim of providing power for BTI and Butwal. But watching electricity's "multiplier effect" at work in Butwal (where training plus power produced jobs, produced industry, and produced more jobs) convinced Hoftun that, if his goal was to build human capacity so as to equip people for good-paying, skilled jobs in an inevitable industrial economy, then Nepal's future depended on its ability to expand sustainable (hydro) power production. Turning its own water into power had to be the foundation of Nepal's hopes of developing an independent, sustainable national economy.

But if these big ideas about industrial development were in Odd Hoftun's mind as he drew up plans for the Andhi Khola project, he did not feel that he could clearly articulate them in the proposals he submitted to UMN.[1] Within UMN Hoftun still faced deep skepticism, and even outright opposition, to the very idea that a mission organization would be involved in promoting industry. "There was *a lot* of resistance in the mission against another hydro project," Hoftun recalled.

---

[1] In his official proposals to the mission, the closest thing I found to a statement by Hoftun of some kind of long-term organic vision for hydropower development is an understated line in the February 1980 "Preliminary Project Description" (Part II, 15) where Hoftun writes: "One could, in several ways, look at the Andhi Khola project as a successor to UMN's power-related activities in Butwal." Hoftun links the project to the past but not to any future vision.

Why should we? [people asked]. We have done a job in Tinau. We have trained some people and it has meant something for the other projects, BTI and so on. So that was justified. But that was also *enough*. It's not our business to do that. They argued that other parties, foreign companies, could come and build that or there could be grants between governments and so on.

Even Hoftun's supporters warned him that if he wanted UMN approval for the Andhi Khola project, it was going to have to be about a lot more than just building up his fledgling hydropower development related companies and producing power for industrialization. For it to have any hope of winning UMN support, the project was going to have to address mission priorities of service to the poor and rural development broadly.

Hoftun was certainly not opposed to the goals of rural development and saw no necessary contradiction between developing hydropower and developing rural economies. In fact, he had long been inspired by similar trends in his native Norway, as well as other massive state-sponsored projects elsewhere such as the Tennessee Valley Authority's rural electrification schemes in the United States of America (USA) during the Great Depression. Bringing power to rural areas could bring huge advances in economic opportunities and quality of life.

The problem is that developing power for *industrial development* is a very different proposition than developing power for *rural development*. In a nutshell, building a power project that services relatively large-scale industrial users within a relatively densely packed urban area is efficient, profitable, and sustainable from a business standpoint. By contrast, building and operating a power plant to service a rural area, where consumers are spread out and likely to use very little power, is terribly inefficient and completely nonsensical businesswise. The costs of building and maintaining distribution infrastructure in rural areas, linking far-flung homes and hamlets, are so high that they can never be part of a sustainable (profitable) private-sector development model. In the past and to this day, virtually all rural electrification (in developed and underdeveloped parts of the world) is heavily state- and/or aid-subsidized. Therefore, building a power plant at Andhi Khola might make sense from a business point of view but not if it was dragged down by a rural electrification project.

For that reason when Hoftun approached Norad (the Norwegian state aid agency) about providing support for the Andhi Khola project, they were "easily persuaded" to provide funds for the power plant in order to support Hoftun's vision of building up Nepali industrial capacity in the hydropower sector. But they were much less enthusiastic about the rural electrification part. "That was too wide and too vague to get funded by Norad," said Hoftun. Therefore, Hoftun faced a dilemma: Norad saw the inefficiencies of rural electrification, but it was the rural electrification component that would (hopefully) make the entire proposal palatable to UMN. Hoftun was conflicted too.

> My vision was industrial development as a necessity for Nepal and for giving people work and income in the future. But the rural development side of it? Not so much, even though I would like to see it spread out across the country.

For the Andhi Khola project to be profitable and therefore self-sustaining it would have to sell bulk power through high-voltage lines to dense consumer areas such as Butwal and Pokhara. But for it to be "approvable" by UMN it was going to have to also service a rural consumer market that was then entirely outside the existing power distribution system. Creating that rural grid in ways that did not break the bank for the rest of the project would require careful planning and innovative design. Hoftun would do this by keeping project costs to an absolute minimum by using local labor, labor-intensive techniques, secondhand refurbished electromechanical equipment from Europe, and simple, cheap rural distribution techniques.[2] By minimizing costs the project would minimize debt from loan financing for construction

---

[2] In a "Preliminary Project Description" dated February 1980 (Part I, 4), Hoftun wrote,

> It is planned that construction methods shall be labor intensive, even if this may result in a longer construction period. There will be a minimum of construction machinery, such as compressors and drilling equipment for blasting, blowers for ventilation, pumps and winches for the vertical shaft, concrete mixers, etc. Rail and trolleys will be used to remove the muck from the tailrace tunnel.

and equipment costs.[3] Freed from the burden of repaying loans and recouping losses from equipment depreciation (old equipment has already depreciated), a project developer can take more time and build a power plant with less cost and more profit potential (Weller, Skeie, and Spare 1992: 729–730). A cheap power plant produces the same product (electricity) as an expensive one. The only difference is that the low-cost project is clear of debt much more quickly and is, therefore, much more profitable (and sustainable) than the high-price project. Hoftun had found a way to combine his commitment to industrial development with the goals of rural development.

But he had not necessarily found a way to convince his colleagues to approve the project. In his February 1979 project proposal Hoftun laid out an "integrated program" of three parts "under one project administration": a 5 MW power plant, a rural electrification scheme, and a simple rural development component that would encourage electricity use in things like small workshops and electric pumps for irrigation. In the UMN board debate went back and forth between supporters and detractors until the project finally came to a vote in April 1979. When the votes were counted the proposal had passed, but barely. "It just squeaked through," recalled one person, in the face of strong, heartfelt opposition. Hoftun may have gotten approval, but it was certainly not a vote of confidence.

Therefore, rather than moving ahead with such a slim margin of support, Hoftun spent another year working up a new version of his Andhi Khola proposal that would try to address his missionary colleagues' concerns. The February 1980 version was now a "multipurpose project" with a much-expanded "general rural development program" built in. In keeping with

---

Otherwise, manual equipment will be used and everything will be as simple as possible.

The reason for this is not only that it saves expenses. It will be the aim to use locally available labor to the maximum extent, in order to maximize the positive effects of the construction work as a factor in the overall rural development program.

[3] More to the point, by minimizing costs the project could avoid costly commercial financing altogether. Project costs would be low enough to finance mainly through grants.

development trends of the day, Andhi Khola was reconceived as a full-blown "integrated rural development" (IRD) project that aimed to touch on many aspects of rural life in a coordinated and mutually reinforcing way. There would be health and sanitation initiatives, literacy and women's empowerment programs, and reforestation and agricultural development schemes. In short, the project would now incorporate almost all kinds of development work that UMN was involved with elsewhere in Nepal.

Perhaps the most significant difference in the new plan was an elaborate irrigation scheme—originally envisioned by the Butwal Power Company (BPC) general manager Balaram Pradhan already in 1978. The large south-facing slope from the ridge at Galyang down to the Kaligandaki River was arid and deforested. Farmers managed to bring in one, and maybe two crops per year but, with no natural water sources, for most of the year the slopes were dry and barren. By modifying the project's design,[4] they could bring Andhi Khola water horizontally through the ridge to a point relatively high on the south-facing slope. From there most of the water would go 250 meters straight down a drop shaft to a powerhouse deep underground. But some of the water would be diverted at the top of the drop shaft for irrigation. In what proved to be, in many ways, a more technically challenging feat than the hydropower project itself, engineers designed an elaborate system of irrigation canals and suspended syphon aqueducts that provided dry-season water to a huge, almost 300-hectare, area. With this water, farmers could double or even triple their output, transforming the area from a food-deficit to food-surplus zone. Hoftun had found a way of combining hydropower development with agricultural development.

Next came the challenges of securing funding, rounding up equipment, building up community relations in the Galyang area, and getting official permission from the Nepal government. Through several installments Norad contributed the bulk of the funding in the form of grants totaling around USD 5 million (NOK 25.4) (Weller, Skeie, and Spare 1992: 727–728; Bakkevig, Hoftun, and Stensby 1996: 14). Norad channeled its money to UMN (via

---

[4] Hoftun had originally proposed a drop shaft just near the Andhi Khola, and then a long tailrace through the ridge. Pradhan's idea was to build a long horizontal headrace through the ridge to where it could also discharge irrigation water.

Hoftun's Norwegian mission association) on the explicit condition that it be used to fund BPC, Himal Hydro, and related Nepal-based design and manufacturing enterprises. Via UMN, Norad money would build the Andhi Khola power plant, and a controlling interest of UMN's shares would go to BPC at the project's completion. For its part, UMN invested USD 2.6 million (in cash, kind, and services) mainly to fund the project's rural development component. In the official agreement with the Nepal government, signed in June 1982, UMN promised to hand over its remaining shares in the project to the government at completion. And finally, the Nepal government committed to providing USD 1.7 million in project financing as well as waiving customs duties and other taxes related to the project (Svalheim 2015: 203). Unlike the Tinau project agreement that forced BPC to hand over the finished product to the government, the Andhi Khola agreement recognized BPC as the finished power plant's owner and operator and even authorized it to directly market electricity to local customers, something never before approved by the Nepal Electricity Authority (NEA). The government would be a minority shareholder but Andhi Khola would be a BPC enterprise in perpetuity.[5]

Even while these arrangements were being hammered out, the project was already underway. In April 1981 American engineer Duane Poppe and Briton Joy Poppe moved into a rented Nepali house in Galyang bazaar. Their job was to establish ties with the community and involve local people in the preparation process. By this time both Joy and Duane had lived in rural parts of Nepal for years, were comfortable with local lifestyles, fluent in Nepali, and quickly made many friends in the area.[6] Joy's initial job was to conduct

---

[5] At the time when UMN negotiated the Andhi Khola agreement with the government, Nepal had no laws regarding the public ownership of water rights. Once those laws came into place (in 1992) all private hydropower development agreements had mandated time limits built into the contracts after which private projects would revert to public ownership. For this reason BPC's more or less permanent ownership of Andhi Khola is a rare, if not unique, case on the Nepal hydropower landscape. Whether BPC can retain Andhi Khola's special status after the project's recent controversial "up-grade" to 9.4 MW is a matter of some debate.

[6] In October 2016 I had the pleasure of traveling to Galyang with Duane Poppe, the first visit he had made in decades. Walking through the bazaar Duane couldn't take five steps without running into someone he knew or who remembered him from their

a baseline survey of local conditions and to find out what people wanted and needed from the planned project. She went door to door throughout the area meeting hundreds of people.

Among other things, Duane helped lay out the project's logistical foundations. He and representatives from Himal Hydro worked with the chief district officer (CDO) to set up a land purchasing committee. Rather than allowing the government to set a price, the committee tried to carefully determine fair market value for the land they needed. When combined with information on the various benefits the project would bring to the area, the land acquisition process went relatively smoothly and without hard feelings. It also fell to Duane Poppe to carry out the project's "environmental impact assessment," or at least what amounted to one in an era before such assessments were required by law. Duane described walking the stretch of river between the proposed Andhi Khola intake and the river's confluence with the Kaligandaki. When he saw that no one was taking irrigation water off the river, he figured the project was a go-ahead. But he was also aware that during the dry season the hydropower project would divert most or even all of the water from this part of the river. He asked local people what they thought of this, given that it might have a negative impact on fish populations. But rather than object, people he spoke with were all for the idea since they figured that the fish trapped in dry-season pools would be easier to catch![7]

In the meantime Hoftun returned to Norway in search of equipment for the Andhi Khola project. Already during the work at Tinau Hoftun had established good relations with the director of the Sorumsand Turbine Factory in Norway.[8] The Sorumsand director had helped Hoftun find suitable used equipment for Tinau and took an active role in helping technicians at BTI properly refurbish the equipment. From then on he continued to be on the lookout for decommissioned equipment in Norway that might be of use in Nepal and in 1977 wrote to Hoftun of three available Pelton turbines

---

childhood. I was impressed with what appeared to be a deep and enduring affection that people had for Duane, and vice versa.

[7] Subsequently, Andhi Khola managers have tried to keep at least some flowing water during the dry season and have installed fish ladders at the weir.

[8] Sorumsand later became a subsidiary of Kvaerner.

with generators that almost perfectly matched the head and volume of the Andhi Khola design. They were free for the taking—if Hoftun could have them dismantled, packed up, and shipped.[9] Again Hoftun mobilized a small army of retired UMNers in Norway, volunteers, and assorted well-wishers: four months, thirty-five people, and 2,700 hours of volunteer labor later, the turbine sets were on their way to Nepal. Still other teams of volunteers dismantled, packed, and shipped still more tons of used equipment to Nepal including large-scale machine tools needed to refurbish turbines, a heavy-duty hoist that would be put to use at the top of the Andhi Khola drop shaft, and even 70 kilometers of used high-tension power line. Some of these volunteers even followed Hoftun to Nepal where they helped train BTI technicians (Svalheim 2015: 204–211). Given the astonishing amount of goodwill poured into it by Norwegians and others, it is hard not to see the Andhi Khola project as a labor of love (which may account for the strong feelings that Hoftun retained for the project for decades to come).

Soon shipment after shipment of equipment started arriving at the BTI facilities in Butwal where it was all carefully brought back to life. I happened to meet two former NEA officials who had been part of an inspection tour to Butwal, led by Odd Hoftun, in the mid-1980s. They both independently described the immense pleasure that Hoftun took in showing them quality used equipment that was destined for a second life in Nepal. Hoftun had a kind of reverence for the old machines (which he suggested were often of higher quality than what was currently available new) and a pride in keeping them in good running condition. In the workshops they saw Nepali technicians skillfully restoring large-scale hydropower equipment. These officials got no sense that Hoftun was pawning off inferior products or depriving Nepal of the most modern machines. Rather, what they saw encapsulated everything Hoftun stood for: stewardship of resources, appropriate technology, technology transfer, and manpower training. They left inspired.

By early 1983 the Hoftun family was back in Nepal and it was time to bring BPC out of its almost three-year hibernation, with Hoftun resuming his position as general manager. By this time Duane and Joy Poppe and other

---

[9] Sorumsand also invited three BTI technicians to come to Norway, work in the turbine factory, and learn how to overhaul and maintain used equipment (Svalheim 2015: 208).

UMNers had created positive relations between the local community and the developers. All project staff (both expat and Nepali) who stayed on site—whether with UMN, BPC, or Himal Hydro—lived in local housing under more or less local conditions, getting to know their landlords and neighbors and interacting with them on a daily basis.[10] Himal Hydro engineers and workers prepared the superstructure above the site where the drop shaft would be built, including erecting the large crane or hoist from Norway. After a false start in 1982, real construction only began in earnest in 1983. Because Himal Hydro was the only construction firm in Nepal with any tunneling experience whatsoever, they were the clear choice for the job. As for design, engineers (both expat and Nepali) from UMN's DCS did the preliminary Andhi Khola work before, in 1982, BPC incorporated the DCS's hydropower design expertise into a special consultancy division within the corporation—BPC Hydroconsult—which then finished the detailed blueprints for the project's headworks and powerhouse (Bakkevig, Hoftun, and Stensby 1996: 14).[11] BPC Hydroconsult also took on the formal role of serving as intermediary and quality control agent between the project developer (BPC) and contractor (Himal Hydro).

Similarly, BEW began overhauling the used electromechanical equipment from Norway before, in May 1985, BEW spun off a new company specializing in electromechanical equipment: Nepal Hydro and Electric Pvt. Ltd. (NHE). The logic was that for the price of having Norwegian professionals overhaul the used equipment *once*, an entire dedicated hydropower workshop could be established (utilizing used equipment from Norway) and Nepali technicians trained to do this kind of work *in perpetuity*. NHE's incorporation was unique because, in addition to having BPC, BEW, and Himal Hydro as combined

---

[10] Eventually UMN built permanent staff residences at the drop-shaft site but expats continued to live in local housing throughout the construction phase.

[11] At least part of the reason for extracting BPC Hydroconsult out of DCS was that some of the UMN expat engineers in DCS were uncomfortable being associated with the commercial/business/industrial implications of the Andhi Khola project. Their consciences were more attuned to developing design solutions that would bring more direct benefits to the poor. Creating a separate hydropower design and consulting service was, therefore, a way of not just focusing expertise, but of quelling dissent. Hydroconsult remained an in-house division of BPC until 2009 when it was registered as an independent corporation, Hydro Consult Engineering Ltd.

majority shareholders (thanks to major grant funding by Norad), the new company also had Norwegian private corporations as minority shareholders. Initially, investment from the turbine conglomerate Sorumsand/Kvaerner represented about a third of the total shares: they would help NHE develop small turbine manufacturing capabilities and help train Nepali technicians in international-standard repair and maintenance work. Some years later the Norwegian company ABB Energi invested an equal amount to set up a new workshop specializing in industrial-scale generator repair and maintenance. In the end the two Norwegian companies held around 47 percent of NHE's shares and BPC another 49 percent, with other Nepali shareholders filling out the remainder.[12] With its new capacities, NHE not only successfully overhauled all of the Andhi Khola project's electromechanical equipment but also produced and installed the complicated high-pressure penstock tubes, valves, and gates that required first-class welding and machining skills.[13]

Hoftun stressed that he saw Andhi Khola as a place where this new small flock of hydropower entities—BPC, BPC Hydroconsult, Himal Hydro, and NHE—could gain "functional experience" in how to carry out their specific roles in an actual project. The companies were still far from operating independently in the "real world" competitive market, but Andhi Khola would be where they took their first baby steps in that direction.

CONSTRUCTION

Digging a vertical drop shaft 250 meters straight down through solid (and sometimes not-so-solid) rock was a completely new experience in Nepal.[14] For the inexperienced people involved in its construction, it was also just plain scary. Himal Hydro staff, like the American civil engineer in charge of construction and Shiva Kumar Sharma, who joined Himal Hydro in 1981 with a fresh degree in mining engineering, had plenty of theoretical

---

[12] In 2004 BPC increased its equity holding in NHE to just over 51 percent.

[13] See Bakkevig, Hoftun, and Stensby (1996: 14); Weller, Skeie, and Spare (1992: 732); and Svalheim (2015: 221).

[14] In fact, Andhi Khola's is still the only vertical drop shaft in existence in Nepal today.

knowledge but no practical experience. Sample boreholes had indicated that they were in the right spot to dig, but how to dig, and especially how to dig safely, was another matter altogether. Just looking down the hole was frightening but worse was the prospect of things falling down the shaft onto the workers. And there was the question of how to communicate between up and down so that things would proceed safely and smoothly. After excavating only 10 meters the supervisor refused to go further out of safety concerns.

Not long thereafter an Australian mining engineer, in Nepal on holiday, happened to hear of the project and dropped by for a visit. Encountering the panicked crew, he described how it was done in Australia—involving whistle signals, strictly assigned roles, and a strict chain of command—and assured them that the same could be done in Nepal. Work resumed and soon became routine with the shaft completed without serious injury. But even so, it was never anything like pleasant work. A Nepali Butwal Power Company (BPC) engineer told me of how people working on the site

> would get scared to death sometimes because you're deep down, just riding in a bucket that was just designed to take out the muck. Guys would go down there drilling, then they'd set the explosives and people would come out. Then they'd blast and you'd go down with the bucket again and load it up with rubble. It was really labor intensive.

As they worked their way down, workers applied 5–10 centimeters of shotcrete[15] to the drop shaft walls to prevent loose rock from careening down the hole and potentially injuring people below. Looking back at his first practical experience after earning his degree, another Nepali BPC engineer assigned to Andhi Khola notes that his main memory of the project was its overall culture of safety and precaution.

> They took safety very seriously and that was unusual in Nepal. It was because of the UMN system. They are mostly people from the developed world and they imported the system, the knowledge, skills; all those

---

[15] As the name suggests, shotcrete is a concrete slurry blown under high pressure onto rock surfaces to keep pieces from breaking off and to minimize water seepage.

things. Because of that we had the opportunity to work with a priority to safety, and also gained knowledge about how to do things safely.

Though no longer with BPC, he still insists on following the safety procedures he first learned at Andhi Khola.

Compared to the drop shaft, excavating the head- and tailrace tunnels were relatively easy (and familiar to those with experience from Tinau): drill, blast, ventilate, remove rubble (using hand-pushed rail trolleys), repeat. Problems came with rock conditions which varied from stable, solid rock to "highly fractured phyllite" that was predisposed to collapse and/or compress. At one such area near the entrance to the headrace a 15-meter chimney collapsed into the tunnel—opening a view up to the blue sky above a rice paddy. No one was hurt and they were able to reinforce that section of the tunnel and fill in the hole. The tailrace and powerhouse had some of the same fractured rock also prone to cave-ins. As at Tinau, they lined the tunnels with concrete and masonry walls. But unlike at Tinau where the rock had been relatively stable, in the Andhi Khola powerhouse and tailrace tunnel they soon saw ominous signs of compression or squeezing as tunnel walls and floors bulged inward up to 30 centimeters in a matter of months. Again geologists assured the engineers that this was normal (for this kind of rock). Just dig things extra wide and deep in those areas and eventually the expansion will stop, they advised. Rock bolts and shotcrete also helped stabilize risky areas, especially in the powerhouse. Because the exit of the tailrace tunnel was far down the slope and far from the nearest road, getting supplies to that workface was a challenge. Rather than build a road, Himal Hydro built two ropeways: one to bring equipment and supplies down from Galyang bazaar to the tunnel mouth and another to carry aggregate from across the Kaligandaki River. Last to be completed were the project's headworks consisting of a 6-meter-high by 60-meter-long dam (Fig. 3.1), two large sediment settling tanks, gates, and valves—all carefully designed to reduce cost but maintain durability (Weller, Skeie, and Spare 1992: 727–730). By the time Himal Hydro finished their work in early 1990, they had become a group of skilled technicians ready for about any tunneling challenge.

Once the tunnels and headworks were completed, it was time for Nepal Hydro and Electric (NHE) to install the project's electromechanical

Figure 3.1 Andhi Khola project headworks below Galyang Bazaar
Source: Photo by Naomi Liechty.

equipment—almost all of which had been either refurbished or fabricated from scratch in NHE's Butwal workshops. This included 250 meters of carefully welded 1.05-meter diameter heavy steel penstock tube and bifurcations (designed to withstand enormous water pressure) to deliver water to the project's double-jet Pelton turbines. Down the same shaft went the three turbine-generator sets. Once installed, these fed power back up the shaft to a large transformer which, in turn, prepared a uniform 33 kilovolts (kV) of electricity for transmission. Technicians opened the valves for trial operations in April 1991 before systems were ready for commercial production three months later in July.

ELECTRIFICATION

One of the unique features of the Andhi Khola project was that, in addition to feeding high-voltage power to Pokhara and Butwal via heavy transmission lines and large-scale towers (designed and built by Nepal Hydro and Electric [NHE]), part of the original agreement with the Nepal government was that the Butwal Power Company (BPC) would be given distribution rights to the districts surrounding the power plant.[16] At that time, the whole area was off the national grid and completely lacking any electrical distribution infrastructure. At Andhi Khola BPC would make money by selling power to Nepal Electricity Authority (NEA) through its high-voltage lines but in order for it to fulfil its commitment to rural electrification, the project needed to find innovative, cost-cutting techniques that would allow it to bring power to rural consumers without overwhelming the company's resources. As one Nepali BPC engineer who worked on the project explained to me, "We were basically designing *systems* for rural electrification that would make it low cost, so that local communities would have ownership of it. Low cost not only for installing,

---

[16] The government's granting of private distribution rights to BPC was unprecedented in 1982 and remains so in Nepal to this day. Instead, today most rural electrification is managed by almost 300 local rural electrification entities (REEs) across Nepal that purchase bulk electricity from NEA and retail it to local customers.

but also for operating." By promoting local "ownership" he meant that the electrification systems needed to be designed in a way that rural people could not only afford, but also manage and maintain. Complicated, high-tech, or dangerous systems would never be sustainable.

Innovation started with distribution lines. Instead of the 11 kV lines usually used for local power distribution, BPC installed simple 1 kV lines. In areas far from roads, these small-gauge lines and relatively lightweight transformers were simple to carry in under animal or human power and they did not require elaborate towers or poles. Lowering the cost of installation would also lower the cost of power to consumers.

Ironically, one of the main components of the total cost of energy for consumers is paying for the labor required to figure out how much power each consumer has used. The cost of paying someone to regularly trek through a rural area reading one isolated meter after another raises the cost of power significantly. To deal with this problem BPC engineers turned to a technique that had been used earlier in Europe: rather than a meter, in each consumer's home they installed a "cut-out" (or "positive thermal coefficient semiconductor"). Rather than measuring the exact amount of power used, these devices allowed users to consume as much power as they wanted in a day, as long as they did not consume above a certain fixed amount of wattage. "Whether you use it two hours a day or twenty hours a day, you still pay the same flat rate," explained a former BPC engineer.

Using "cut-outs" helped improve the system's efficiency in several ways. First, the units themselves were essentially free because they were used equipment donated in bulk from Norway. Second, cut-outs turned power consumption into something like a subscriber service: because there were no meters and rates were fixed, there was no need for a meter reader. Instead, consumers formed local user groups and appointed one person to collect fixed fees on a monthly basis, in return for a small percentage of the fees. For BPC this meant that, rather than billing 150 individual consumers, they dealt with one local representative who came to their office and paid for 150 households. And third, cut-outs were a very effective way of controlling peak load and spreading load out over the day—instead of the typical pattern in domestic consumer markets of demand spikes in the morning and evening. In the absence of industry, hydropower (which produces the same amount of electricity regardless of demand fluctuations) is often underutilized throughout the day and late at night.

One of the Andhi Khola project's long-term goals was to promote local industry and other power usage (for things like irrigation pumps, ropeway systems,[17] and so on) so as to even out demand for the new power supply. But in the meantime, engineers from BPC and the United Mission to Nepal's (UMN's) Development and Consulting Services (DCS) tried to promote systems and products that would help consumers get the most out of their power supply and spread demand throughout the day and night. Most notable was the *bijuli dekchi*, or electric cooking pot, a kind of low-wattage slow cooker made of simple local materials. The majority of domestic energy use in rural Nepal was for cooking, with heat coming mainly from wood and animal dung. With a basic 250-watt service, a cook could use the *bijuli dekchi* to heat water during the night or afternoon; the water, in turn, could then be used to speed up cooking during the morning and evening. At a time when deforestation (for fuel wood) and respiratory problems due to indoor cooking fires were both major concerns, rural electrification promised to improve both the environment and health conditions.

Another benefit of the extremely low-voltage distribution system was safety. Especially once the power was stepped down to domestic-use levels, it was much less of an electrocution threat and community members could learn how to handle it without elaborate training or equipment. The very low-voltage lines also meant that you could install them in typical bamboo and thatch village homes without a high risk of fire or electrocution. In fact, BPC engineers devised simple house-wiring kits consisting of lines with fixed lightbulb sockets that virtually anyone could install in minutes with very simple tools.

When taken together, the numerous innovations and cost-cutting measures that went into the Andhi Khola project's rural electrification scheme were enough to attract the attention of international development experts. One such was Allen Inversin of the US-based National Rural Electrification Cooperative Association, who wrote a lengthy report on the project and

---

[17] For more on ropeway systems as appropriate technology in mountainous regions, see Gyawali, Dixit, and Upadhya (2004); Upadhya (2017).

heaped praise on its creative and practical approaches.[18] But the people BPC and DCS engineers most hoped to impress were NEA officials. "What Hoftun and all of us hoped was that it would make a strong case to NEA to consider this as an alternative to what they were doing," said one former BPC staffer. They hoped to have shown that their "simpler is better," "appropriate technology" approach was a viable option for spreading electricity throughout rural Nepal, one that avoided the high construction and operation costs of standard grid coverage.

But the careful philosophy that guided the project's initial rural electrification expansion did not last. Eventually there were problems with the cut-out system. All of the units had come used from Norway and when the supply ran out, and spare parts were no longer available, they had to be discontinued. Also, consumers soon found ways to tamper with them. A carefully placed magnet would prevent the cut-out from cutting out at high usage! The same was true of other simple two-tiered meters that had been brought used (and free) from Europe. Once the supply dried up and they were no longer even manufactured, they were out of the picture. Even the project's simple home lighting systems met resistance. Many rural households, unable to pay the 70 Nepali rupees (NPR) per month fee for the basic 250-watt service, were soon disconnected. People found that running three lightbulbs, while providing better light, was more expensive than using three kerosene lanterns (Poppe 1993: 21). The extremely cash-poor rural economy meant that many people were still not ready for even the most basic, cost-reduced electrification.

The fate of the *bijuli dekchi* slow cooker is also illustrative. It heated water, but *slowly*—usually four–six hours to bring it to the point where it could be used for cooking. Using it effectively meant changing a family's cooking behavior, something that most were not eager to do. The cookers also found resistance from Brahmin households who adhered to strict rules about food preparation and purity. For them water heated overnight was already day-old and therefore *baasi*—or "stale" and thus ritually impure—and had to be thrown out. The final nail in the slow cooker's coffin was the arrival of

---

[18] It was largely on the basis of this report (Inversin 1994) that the Andhi Khola project later received the prestigious Blue Planet Prize.

Chinese rice cookers. They were cheaper, more attractive, and faster: the DCS could not compete.

As more and more roads penetrated rural areas another of BPC's carefully thought-out cost reduction techniques became irrelevant. The cost and difficulty of hauling in heavy 11 kV lines and transformers over mountain footpaths had initially made 1 kV lines the only viable option. But as more and more villages and hamlets had road access, it was relatively cheap and easy to haul in even heavy equipment. Heavier transmission lines meant that it was now possible to use "standard technology" (for example, standard distribution voltages and electricity meters) rather than using "appropriate technology" out of necessity. By the time BPC was nationalized (became government owned, see Chapter 4) in 1996, there was no point in challenging "the establishment" (commodity production approach). BPC had *become* the establishment.

IRRIGATION

Of all the elements of the larger Andhi Khola project, the irrigation scheme was one of the most technically complicated, socially ambitious, and, in the long run, most successful. Even just visually the results are dramatic. When I visited the area in late 2016 it was unrecognizable from pictures taken in the 1970s that show a vast, empty, deforested, south-facing slope stretching from Galyang bazaar down to the Kaligandaki. Today forests of 10- to 15-meter-tall trees and underbrush stand between lush green terraces stepping down the hillside.

The same innovative spirit that went into the rural electrification program also went into the irrigation scheme with creative approaches to both technological and social challenges. The engineering challenge was to distribute the 0.6 cubic meters per second of water across the steep and geologically unstable mountainside so as to cover as much area as possible. Rather than simply hugging the mountain's contours, Butwal Power Company (BPC) engineers designed a 7-kilometer system of horizontal canals (across more stable portions) and suspended syphon pipelines across the area's many loose, landslide-prone ravines. The longest of these suspended pipelines swoops 110 meters down and up across the Kaligandaki River to irrigate land on the other side. Another challenge was how to fairly divert water into secondary canals without resorting to complicated valves or gages. The answer was small horizontal weirs (perfectly flat spillways or mini dams) with simple calibrated gates such that, for example, to divert one quarter of the flow one moves a board to the one-quarter mark on the weir. In this way the weirs do not determine the *amount* of water flowing into secondary canals but the *percentage* of the total flowing through the main canal at any one moment. Where water flows down steep secondary canals engineers designed special baffling devices to slow the flow and reduce wear on the canals. A Nepali engineer I spoke with who worked on the irrigation design called it a "very extensive" project and "very interesting from an engineering point of view."

Even before the engineering work began United Mission to Nepal (UMN) staff were working with local residents to determine how the irrigation project would be implemented. This water users group came to be known by its English acronym—AKWUA (sounds like "aqua") for Andhi Khola Water Users Association, representing over 1,300 local households. Comprised of members from both landowning and landless families, AKWUA determined that water would be allocated not according to who owned land but to those who contributed labor or money to building the infrastructure. That meant that even landless people could, in return for labor inputs, receive water rights that they could then either sell or rent. Out of a total of 25,000 available project shares, anyone contributing five days of labor, or NPR 165, received 1 share. The association also limited the number of shares that any one household could own to prevent anyone from hoarding (Weller, Skeie, and Spare 1992: 731).

Perhaps even more innovative was the land redistribution policy that AKWUA negotiated whereby landowners over a certain size were required to give up 5 percent of their land in return for getting water.[19] The logic was that with irrigation landowners would see *at least* a 50 percent increase in production

---

[19] Writing in the early 1990s, Weller, Skeie, and Spare (1992: 731) put the figure at 10 percent for land required to be handed over. But people I spoke with said that in the long run the project managed to redistribute closer to only 5 percent of the designated land.

(from two crops to three) in comparison to which giving up 5 percent of one's land was negligible. The land thus freed up would go to landless families. Duane Poppe estimated that around forty families received 0.25-hectare plots, the amount taken to be the standard subsistence-level family farm in Nepal.

Eventually the Andhi Khola irrigation scheme watered some 282 hectares of land supporting around 800 households (Bhandari and Pradhan 2006) but it took a long time to reach those numbers. As of 1993, two years after the Andhi Khola power plant had commenced operation, the irrigation project was only one-fifth complete, fewer and fewer people were showing up for volunteer work, and "One sensed that people were losing faith in [the project's] ability to deliver their promises" (Poppe 1993: 21). But eventually local people working with UMN engineers did finish the complicated suspended syphon "bridge crossings" finally making water flow across the dry mountainside. Today AKWUA remains a very effective organization with users completely in control of the system's operations and maintenance. Unlike most government- and/or aid-instituted irrigation projects in Nepal—in which, when they break down, people wait for outsiders to come and fix them—AKWUA had local ownership built right into its fabric from the beginning making it a truly sustainable community organization.

CONCLUSION: CONTRADICTIONS AND ACCOMPLISHMENTS

In retrospect it is hard to see the Andhi Khola project as anything but a success. In terms of power generation, the extreme efforts that went into keeping costs low (grant funding, used equipment, labor intensive construction methods) meant that the project very quickly paid off its debts and began generating pure profits for the Butwal Power Company (BPC). The project's ratio of (low) investment input to (high) kilowatt output was unsurpassed in Nepal, making Andhi Khola the country's most profitable power plant (at least until controversial upgrades decades later changed that equation—see Chapter 8).

On the social side of the story, AKWUA stands as testimony to the Andhi Khola project's ability to seriously engage the community and have local people literally invest in, and therefore own, their own futures. Although

AKWUA and the project's other more rural development-oriented aspects were distinct from BPC's technological undertaking (the power plant and rural electrification), there is no question that BPC—as a local commercial power provider—benefitted from the large degree of community "buy in" and plain goodwill that the project as a whole produced. (Just how much becomes plain in comparison with the Jhimruk project discussed in the next chapter.) With its large-scale "integrated rural development" (IRD) component added on, the Andhi Khola project was able to generate a lot more than just power (Nafziger 1990).

But even at Andhi Khola there were disappointments, tensions, and contradictions that weighed on the project. As discussed earlier, many of the project's engineering innovations did not withstand the test of time. And at the broadest level the project failed to significantly transform the local social structure by uplifting "the poorest of the poor"—even if it did bring real advances in many development indicators including access to clean drinking water, use of toilets, female adult literacy, reforestation, and others (Poppe 1993).

Probably the most notable contradiction at Andhi Khola, one largely invisible to local people, was the fault line present in the project from its very inception: between the commercial or industrial priorities of the hydel project and the grassroots social advancement priorities of the rural development project. Hoftun had beefed up the latter in order to win support for the former but the two never really coalesced into one project. When I asked Nepali professional staff who had worked at Andhi Khola whether there were tensions between expats and Nepalis on the project, the answer was "not much." A few foreigners allowed their "missionary impulse" to impinge on their relations with Nepali coworkers in unwelcome ways. But for the most part the strains between the two groups were minor, having to do with imperfect communication and differing cultural expectations. Nepali and foreign staff lived under the same local conditions (housing, and so on) and, from the stories I was told, generally enjoyed very good relations.

Interestingly, Nepalis reported that the *real* conflicts were not between them and foreigners but between the foreigners themselves. One United Mission to Nepal (UMN) expat I spoke with agreed.

The most tension was between the project and the so-called "companies." It was the [rural development people] versus the companies. I mean there wasn't a lot of outright hostility but there was definitely tension between the project and the companies because the project had more community outreach and a kind of go-slow, steady-but-sure approach. Whereas the companies had deadlines to meet, contractual obligations to meet, and so on.

He noted that BPC, Himal Hydro, and Nepal Hydro and Electric (NHE) staffers were "there for business" whereas the other project members had more of a service orientation. From my own conversations with Nepalis and expats working for "the companies," I know that they too were clearly inspired by a sense of service and even mission—to promote economic development, job creation, and national independence. But these two visions were rarely visible across the divide.

In fact, it was during the Andhi Khola years that the conflict between Hoftun's pro-business and other UMNers' antibusiness sentiments came to a high point. Through the mid-1980s, with the power project's construction under full swing and Hoftun busy incorporating NHE with the help of major Norwegian companies, those within UMN opposed to Hoftun's initiatives were also gearing up. The anti-commercial wing of the mission finally convinced UMN leadership to commission an evaluation of Hoftun's work by researchers from Britain's Cranfield University, an influential development or aid program that had, in fact, trained many of those UMNers who now sought Cranfield's assessment. Named for its author, the 1988 Grierson Report condemned UMN's industrial development work in no uncertain terms.

> Grierson categorically declared that the mission's shareholding companies were organized with only economic profit in mind. They were thus disgraceful and irreconcilable with the principles of development aid. He [Grierson] researched the commercial interests and connections to business in Norway and found them to be suspicious. He also raised questions about the project's accounts. (Svalheim 2015: 238)

In response, Hoftun hired Ratna Sansar Shrestha, then UMN's external auditor, to thoroughly review the companies' collective books. Shrestha found no signs of irregularities, wrongdoings, or unethical behavior—but even this hardly exonerated Hoftun in the eyes of his detractors, for whom *any* capitalist commercial enterprise was, by definition, unethical. The ethical industrialization that Odd Hoftun, Balaram Pradhan, and others saw as the heart of BPC's mission in Nepal represented a reconciliation—between business and service, market growth and "community outreach," profit taking and reinvestment—that many simply could not make.

Yet this dis-ease with commercial, market-based approaches was much more likely to afflict foreigners than Nepalis within the larger UMN family of organizations to which BPC belonged. The Nepalis who embraced Hoftun's vision as a means to achieve their own vision for Nepali national advancement saw industrialization as inevitable, at least if Nepal was ever to achieve anything like economic independence. In this shared vision hydropower was the key to job creation and maybe even prosperity. I found it interesting that even among the Nepalis in the Galyang and surrounding areas most immediately affected by the larger Andhi Khola project, they saw no contradiction between its integrated rural development (IRD) and hydropower components.[20] In fact, it was the combination of the two that produced so much goodwill toward the project as a whole, goodwill that BPC's Andhi Khola power plant still benefits from today in the form of amicable community relations.

---

[20] The one exception to this lack of conflict, in local perception, between the hydropower and community development dimensions of the Andhi Khola project, might be in regard to water allocation for irrigation. Here, by definition, water emerging from the headrace that goes for irrigation does not go for power generation. A UMN staff member who worked on the project described how, in advance, AKWUA had discussed and agreed upon the amount of water that would be drawn off for irrigation. But when the irrigation system was finally up and running, local users realized that they could use *more* water.

4

JHIMRUK

While the Andhi Khola project is a veritable model of successful community-based hydropower development, the Butwal Power Company's (BPC's) next project—on the Jhimruk Khola in Pyuthan District—was a much more ambiguous achievement. Andhi Khola received international recognition for its technical and social innovations even while it generated power for a remarkably low price in terms of initial investment. The Jhimruk project too admirably succeeded in achieving its *technical* goals of boosting levels of professionalism for BPC and its collaborating companies while building a high-quality power plant on time and within budget. But whereas the Andhi Khola project had carefully worked to cultivate good relations with the local community and integrate community goals and concerns into the larger project from start to finish, at Jhimruk community relations started off on the wrong foot and deteriorated from there. While development experts now often point to Andhi Khola as an exemplary case study, the Jhimruk project is more likely to serve as the opposite—an example of what can go wrong without careful consideration of local conditions—both social and environmental. Even now, more than two decades after its commissioning, the Jhimruk project is the target of local hostility and mired in political controversy (Kunwar 2016), as well as unforeseen environmental problems.

At the root of this contrast between Andhi Khola and Jhimruk is the fundamental difference in how the projects were conceived and implemented. Whereas Odd Hoftun and the United Mission to Nepal (UMN) had initiated Andhi Khola (like Tinau before it) as part of Hoftun's larger vision of promoting industrial development through hydropower and skilled Nepali labor, Jhimruk was a *commissioned* project taken on by BPC at the request of the Nepal government. Because BPC was officially just fulfilling a government contract, they expected that the government itself would assume full responsibility for the finished project. Rather than laying the groundwork for a long-term relationship with the community as they had done at Andhi Khola, at Jhimruk BPC saw its role as almost entirely technical: it would deliver a product (a power plant) to a consumer (the Nepal government) and then walk away. As for assessing the project's social and environmental implications and impacts, BPC left those problems to the government which, in turn, failed to address them. By the time anyone was aware of the mess the project had gotten itself into, it was already too late and—as things turned out—BPC was left holding the bag.

Already in the early 1980s a British engineer-geologist advising the Nepal government alerted Hoftun to a promising site in the Jhimruk Valley about 120 kilometers northwest of Butwal where, much as at Andhi Khola, two rivers running parallel but at different heights and separated by a relatively narrow ridge offered promising hydel potential. Nepali authorities had already proposed a tunnel-based hydropower project with a 200-meter drop and were actively looking for developers and funding. At 12 megawatts (MW) the Jhimruk project would be more than twice the size of Andhi Khola but the Nepal government encouraged BPC and its allied companies to consider taking on the job.

In 1984 construction at Andhi Khola had literally only just begun but the Jhimruk project was an attractive prospect. Hoftun was well aware of the lengthy lead time needed for any hydroelectric project and, if he wanted to keep on building Nepal's indigenous power production capacity, he needed to have another project lined up that would commence about the time Andhi Khola was winding down. Already in 1984 Hoftun got encouraging signals from Norwegian aid officials about the prospects of a government-to-government grant for the project and began work on a detailed funding proposal that the government of Nepal officially submitted to the government of Norway in March 1987. In the proposal Hoftun identified three main goals for the project: to provide power to the national grid, to electrify the local area, and—most importantly for Hoftun—to provide rigorous professional experience for BPC and its spin-off companies Himal Hydro and Nepal Hydro and Electric (NHE). By July 1987 Nepal Electricity Authority (NEA) engineers had completed a very basic (and flawed) feasibility study at the Jhimruk site and then hired BPC's Hydroconsult engineering wing to develop a detailed technical project proposal, which they finished in October 1987 (Bakkevig, Hoftun, and Stensby 1996: 14–15). In May 1988 the Norwegians agreed to fund the entire USD 19 million project on the condition that it be carried out by Nepali companies. The final formal agreement was in place by February 1989 after which planning, design, and bidding for materials took up the rest of the year before construction finally began in 1990. Just as Hoftun had hoped, Himal Hydro crews were able to shift almost seamlessly from the completed civil works at Andhi Khola to the new project at Jhimruk.[1]

In this chapter I will first lay out the story of the Jhimruk's construction before doubling back to consider the social and environmental problems that soon enveloped the project.

CONSTRUCTION

From the outset Odd Hoftun viewed the Jhimruk project as an elaborate training exercise. From equipment operators to overseers to engineers to managers, Jhimruk would provide intense, practical, on-the-job training for hundreds of Nepalis who would leave the project with skills needed to assume increased responsibility and independent leadership roles. At all levels many of the Nepali staff had previous experience at Andhi Khola and many were also tradesmen trained by the Butwal Technical Institute (BTI). But whereas Hoftun had earlier been willing to compromise in terms of technical standards and procedures in recognition of still developing local capacity, at Jhimruk he aimed to raise the quality of design and construction to true international standards. To do this, from the beginning the project incorporated elaborate partnerships with Norwegian institutions and corporations with foreign advisors providing technical advice to their Nepali counterparts. Previously Nepalis had worked with UMN expats but Jhimruk was the first project to formally incorporate foreign technical advisors assisting in everything from

---

[1] In addition to my own archival and interview data, this paragraph and the rest of this chapter draws from Bakkevig, Hoftun, and Stensby (1996) and Svalheim (2015).

hydrological modeling to turbine design to high-tech welding to business management. Hoftun saw Jhimruk as an opportunity to bring Nepali skills and experience up to a level at which they could compete with international hydroelectric development entities.

But competing at an international level was about much more than developing individual professional capacities: even more importantly it was about raising *institutional corporate capacity*. Hoftun was intensely aware that, ultimately, it was as stand-alone corporate entities that BPC, Himal Hydro, and NHE would succeed or fail in the complicated and cut-throat international market for hydroelectric development in Nepal. Up to then BPC and its allied companies had been technically distinct corporate entities but in reality were still all like chicks in the same cozy nest provided by Hoftun, UMN, Norad, and the fact that they had no competitors in Nepal. Although their roles were distinct, each company dealt with the other in a more or less informal, unstructured, ad hoc way. Hoftun knew that in a "real world" business environment all this would have to change. Each company would have to deal with others on a formal contractual basis, under which they would each be responsible for protecting their own interests, maintaining profitability, and sticking to agreed-upon schedules. Of course the Jhimruk project was far from a "real world" situation: the different corporations involved were closely related, some companies held shares in the others, and the project was donor funded, not commercially driven (and therefore devoid of major risks and pressures attendant with any loan-financed operation). Nevertheless Hoftun wanted Jhimruk to be a training exercise to help develop "independent and self-sustaining indigenous institutions" (Bakkevig, Hoftun, and Stensby 1996: 55).

Much of what was expected of them was completely new for the Nepali corporations involved. For example, if BPC was to be the project's official developer and owner, Himal Hydro would have to formally bid for a construction contract. For the first time, Himal Hydro had to come up with a full cost proposal (for materials, equipment, and labor) for a complicated multiyear project. Determining and meeting these costs is a matter of life and death for an independent company, and something they had to get used to handling. In the end Himal Hydro's contract with BPC had forgiveness clauses guaranteeing that if their cost estimates were too low or if they fell behind in the construction schedule, they would not lose money. But even so, there was great pressure on Himal Hydro to carefully track income, expenses, and reserves. Jhimruk would be an elaborate "test run" for Himal Hydro before releasing it on its own.

BPC's relations with NHE were much more complicated. To begin with, because BPC was a major shareholder in NHE, there was often confusion and conflict over management decisions. Even though BPC officially tried to keep its dealings with NHE on a formal contractual basis, NHE often felt that BPC used its power as owner to sway decisions. For its part, BPC complained that NHE showed little initiative and expected BPC to do its work for it. This was partly due to the complicated nature of the project's electromechanical contracts. In a "real world" situation, NHE would have contracted with Norwegian equipment suppliers on its own. But, because of grant stipulations and in order to keep costs down, in this case BPC took responsibility for importing foreign goods, leaving NHE only the job of manufacturing and installing—thereby leaving inevitable tensions and uncertainties between the two entities. One unique feature of the Jhimruk project's administrative structuring was the appointment of a "quality auditor"—a very senior and respected Norwegian engineer whose role was to serve as a kind of referee between the various Nepali corporations. He helped the companies comply with their contractual commitments and to arbitrate when disagreements arose among the contracting parties. In effect this "auditor" was like a business coach trusted by all sides to help them individually and collectively learn the rules of the game and preparing them for the future when they would be on their own.

Technically, funding for Jhimruk went through a complicated chain of entities (from Norad, to Hoftun's Norwegian mission organization, to UMN, and finally to BPC[2]) but in practice Norad money went straight to BPC which, in turn, entered into numerous contracts with suppliers in Norway (a role that, technically, NHE should have taken). Even though the Norad agreement did not stipulate that Norwegian equipment had to be

---

[2] In fact, UMN transferred project funds to BPC by purchasing BPC shares, which UMN held until transferring them to the Nepal government when BPC was nationalized in 1996.

used, Norwegian project directors felt that Norwegian money should go for Norwegian equipment, even if it was significantly more expensive than some alternatives.[3]

Dealing with Sorumsand/Kvaerner (for turbines) and ABB Energi (for electrical components) also made sense because these companies had already begun working with BPC on the Andhi Khola project. Although both companies officially viewed their activities in Nepal as part of a long-range business strategy (seeing Nepal as an important market for their products in the future), in fact their motivations went beyond strictly business. Individual staff from the Norwegian companies already had relationships with Nepali colleagues in NHE and were personally interested in seeing them succeed. Members of their respective corporate boards were also sympathetic to Hoftun's vision and were willing to contribute to its success. As a result, both companies agreed to work with BPC on a simple contractual basis without costly legal complexities or security demands. And, more importantly, even though the prices were high, both companies agreed to work closely with their Nepali counterparts in NHE. For its part Sorumsand/Kvaerner—in return for payment as though the entire turbine assemblies would be produced in Norway—agreed to provide key proprietary turbine components (guide blades, "runners," and electronic governors) and then actually send its personnel to Butwal to train skilled workers, supervise production, and guarantee quality control as *Nepalis* manufactured the bulky turbine casings. In effect the relatively high price paid to Sorumsand/Kvaerner went not just for high-quality turbines but also for *the transfer of skills and technology* to NHE which emerged much more competitive and competent than before. Legally the deal could have been disastrous for BPC (there were no formal, legal stipulations to guarantee a desired outcome) but the deal, like that with ABB, worked because of the trust and goodwill that all parties brought to the table. Although certainly

---

[3] Unlike at Andhi Khola where Hoftun had insisted on using refurbished equipment from Europe (to keep costs down and to provide NHE technicians with experience in equipment maintenance), at Jhimruk he built new equipment purchases into the budget with the goal of giving the Nepali companies experience in relating directly with foreign equipment manufacturers.

out to make money, the Norwegian companies entered the deal in the spirit of international assistance and cooperation.

Building on the skilled Nepali manpower already developed at Andhi Khola, at Jhimruk the majority of staff members, including some in top management positions, were Nepali. While Hoftun continued as general manager of BPC, much of the logistical and administrative heavy lifting went to BPC's Jhimruk project manager, Balaram Pradhan. After serving as BPC's general manager for much of the 1970s, Pradhan had spent most of the 1980s at the head of his own private company. But in 1991 Hoftun happened to meet Pradhan on the street in Kathmandu. "By then BPC was just starting up the Jhimruk project and I was in trouble because the project manager had to return to Norway," Hoftun recalled in 2015. "I asked Balaram if he would be willing to take it over. He agreed, and we have been working together almost continuously ever since." As project manager, it fell to Pradhan to manage complicated import and payment formalities, general logistics, and to work through BPC's formal contractual ties with Himal Hydro and NHE. In addition to being formally responsible for overseeing work done by Himal Hydro and NHE, within BPC Pradhan supervised a staff of twenty-five to thirty engineers, overseers, transmission line survey crews, and support workers, only two of whom (a resident engineer and the "quality auditor") were foreigners (Bakkevig, Hoftun, and Stensby 1996: 21). Himal Hydro and NHE had somewhat higher ratios of expats to Nepalis with UMN-seconded senior engineers each working with two or three Nepali junior engineers—people with varying degrees of formal training but very little on-the-job experience. Overall at Jhimruk the vast majority of manpower at every level came from Nepalis.

Another of Hoftun's goals was to have Jhimruk be an "in house" project with the Nepali companies responsible for (or at least very closely involved with) every phase of the project from start to finish. This meant that, for example, even though BPC contracted with Norwegian equipment suppliers, it was still completely up to BPC's in-house Hydroconsult engineering wing to determine the project's technical design and specifications. BPC engineers had to determine the project's exact hydraulic capacities, exactly how the hydraulic power would be delivered to the turbines, exactly what electromechanical equipment was needed and how it would be installed in the powerhouse, and then control the quality of the work. These ultimate design and management decisions are crucial to a project's success and it was important that BPC, drawing on its previous experience, be able to take this responsibility without relying on outside consultants. Having a proven track record in this regard would allow the company to stand on its own in a competitive market.

From an engineering standpoint, Jhimruk was a fairly uncomplicated project. The site required only a 1-kilometer headrace tunnel and a 380-meter inclined pressure shaft to produce 205 meters of head. The powerhouse was above ground (near where the water drained into the Madhi Khola) and equipped with three 4 MW Francis turbines and three 5 megavolt-ampere (MVA) generators. BPC also erected over 200 kilometers of high-voltage transmission lines as part of the project.[4]

The project's headworks proved to be by far the most challenging part of the overall design. The Nepal government's initial plan had placed the intake dam in a narrow rocky gorge where the structure would be most secure and the least challenging to build. But Hydroconsult engineers pointed out that by moving the intake structure farther upstream on the Jhimruk Khola they could both minimize the amount of tunneling needed and increase the project's total head. The problem was that the ideal location in terms of generating capacity was anything but ideal in other respects. Building the intake on a 250-meter-wide sandy flood plain was a huge challenge in that shifting river channels and monsoon floods could easily wash away any but the most carefully planned structures.

When it became clear that physical hydraulic modeling would be necessary to design and test the Jhimruk headworks, in 1988 BPC Hydroconsult brokered an agreement between Nepal's Tribhuvan University (TU) Institute of Engineering and the Norwegian Institute of Technology to jointly build a hydraulic modeling laboratory on TU's Patan engineering campus. Funded by a special grant from Norad, the program provided training to Nepali students by experienced Norwegian engineers.[5] After testing numerous

---

[4] See Bakkevig Hoftun, and Stensby (1996: 6) for more technical details.

[5] Svalheim (2015: 253–255) explains how this agreement also benefitted the Norwegians. By the 1980s virtually every potential hydropower site in Norway had been developed,

models, Hydroconsult engineers (working with Norwegian experts) settled on a design featuring a sweeping 205-meter-long (by 3-meter-high) diversion dam with a wide concrete apron to settle floodwaters such that they would not undermine the structure. The design also required 1.6 kilometers of upstream "river training": forcing the river into a permanent channel by lining its banks with thousands of large, galvanized wire mesh containers ("gabions") filled with rocks. With further Norad funding, the Patan hydraulic modeling facility later became Hydro Lab Pvt. Ltd., an independent commercial and research entity that remains one of the most important facilities of its kind in Asia.

The job of constructing the "civil works" (the project's physical structures and tunnels) fell to Himal Hydro. A project manager oversaw four senior engineers (in charge of headworks, tunnels, powerhouse, and transmission lines respectively), each of whom supervised three or four junior engineers. Himal Hydro employed 30–40 technical and administrative staff, including support personnel and 7 or 8 UMN expats. These in turn supervised a pool of construction workers that averaged around 700 and came close to 1,000 at the project's peak in May 1992. Most of the skilled and semi-skilled workers came from the Andhi Khola project, with the remaining unskilled labor recruited from the local area.

For the Nepali engineers who had cut their teeth with Himal Hydro at Andhi Khola, Jhimruk was a welcome change in that, for the first time, they were able to use modern tunneling and construction techniques and equipment. Whereas at Andhi Khola Hoftun had made a virtue of using the simplest, most labor-intensive techniques—to save money but also to demonstrate to Nepalis what they could do with basic equipment and few resources—at Jhimruk the Norad funding was, as one Nepali engineer remembered, "very generous." To simulate a "real world" construction contract (where commercial financing would demand rapid project completion),

---

leaving no opportunities for Norwegian engineering students to get practical experience. As a result university enrollments in hydropower-related studies had crashed, leaving professors to fear for the future of their programs. In need of "a new hydropower country," Norwegians jumped at the opportunity to cooperate with Nepal.

Hoftun knew that fast and efficient construction techniques were necessary if Himal Hydro was to meet its planned four-year deadline.

Even so, when the time came to actually allocate money for equipment, Hoftun's impecunious habits proved hard to overcome. Even with plenty of money, buying expensive equipment violated some of Hoftun's most basic instincts about appropriate technology, dependence on foreign suppliers, and the value of manual labor. One senior Nepali engineer remembered being part of

> a big discussion about what kind of equipment to buy. Odd was very unhappy to invest money in big excavators or loaders, that kind of equipment. But we could persuade him! In our meetings with Odd Hoftun we [top Nepali and expat engineers] had to put a lot of pressure on him to allocate money. But in the end we worked out a very good model.

The compromise was that, since the Norad grant money was officially channeled through the Nepal government, the government could officially buy and own the equipment and then hand it over to Himal Hydro at the job's completion in return for ownership shares in the company. "So Himal Hydro got the equipment, and the government got more shares in Himal Hydro. It was a good model and Himal Hydro ultimately got good equipment. But Odd wasn't very happy!" Yet, in fact, aside from the heavy Caterpillar earthmoving machines, much of the equipment at Jhimruk had been used earlier at Andhi Khola, or was imported used but in good condition from Europe (cranes, pumps, fans, rock drills, and so on).

The Jhimruk headworks were unprecedented in Nepal for their innovative design but so were other aspects of the project. Fresh from their work at Andhi Khola, Himal Hydro tunneling crews had no real trouble excavating the headrace tunnel using manual Atlas Copco pneumatic rock drills that could bore holes 1.5 meters deep. (One Nepali engineer recalled, "The Scandinavians would say, 'Just one person per drill.' But these were such heavy things that for Nepalis we needed three people: two behind and one holding at the top!") With explosives and hand clearing of rubble they averaged around 10 meters per week.

Much more innovative was the project's inclined pressure shaft, the first of its kind in Nepal. Unlike at Andhi Khola where the vertical drop shaft had to be excavated—slowly and laboriously—from the top down, at Jhimruk the drop shaft was at a 45-degree angle and excavated from the bottom up. Himal Hydro used a device called an Alimak Raise Climber featuring a rail-based platform set at 45 degrees to the shaft so that drillers could stand on a flat surface at the rock face as they worked their way up the incline shaft. Once holes are drilled, workers evacuated, and the rock blasted, the beauty of this system was that the rubble simply fell to the bottom of the shaft. Gravity did the work that many humans had done at Andhi Khola.

Tunneling took a year and a half to complete but the most time-consuming part of the civil works was the intake structure. Even with heavy earth-moving equipment it was a big job made more complicated by the fact that construction could only occur during the dry season when the Jhimruk Khola was at a low ebb. Overall the project required excavating some 130,000 cubic meters of earth and rubble, and installing 11,000 cubic meters of concrete. Following international construction standards, technicians tested all of the concrete mixes for quality and hardness before being approved for permanent application. This often meant multiple trial mixes before pouring could begin. Initially Himal Hydro imported Indonesian and Malaysian cement but eventually found an Indian supplier that met their standards and saved them money.

Simultaneously with the civil works construction, Himal Hydro engineers and technicians were busy surveying, constructing, and installing major powerlines that would be ready to link the Jhimruk power plant with consumers as soon as the generators were up and running. This included a main 132 kilovolts (kV) line running south to the Tarai. Running through rugged hills, the lines had to span gaps as long as 900 meters. This required large-scale towers—some up to 25 meters tall—each of which had to be constructed on site with parts designed and prefabricated by NHE. What's more, each part had to be small enough to be carried by a human porter through rough terrain. Himal Hydro also built 162 kilometers of 33 kV powerlines through rugged roadless areas as part of the Jhimruk project's rural electrification scheme. When it was all done, Himal Hydro had become "the best equipped and most experienced transmission line contractor in the country" (Bakkevig, Hoftun, and Stensby 1996: 40).

Perhaps even more than it was for BPC or Himal Hydro, the Jhimruk project was especially important for upgrading the professional competence of NHE. Once Himal Hydro had completed the inclined pressure shaft, it was up to NHE technicians to lower prefabricated sections of a 1.5-meter-diameter steel penstock pipe into the shaft, carefully weld them together, and encase everything securely in concrete. In addition to the pipes themselves, NHE also manufactured all the huge steel gates, valves, and bifurcations (as well as the transmission towers and poles mentioned earlier) needed for the project.

But by far the most challenging job for NHE was manufacturing the three 4 MW Francis turbines under license from Sorumsand/Kvaerner. As part of an earlier (non-BPC) contract, NHE had manufactured one small (250 kilowatt [kW]) Francis turbine (according to Sorumsand/Kvaerner design specifications) that it successfully installed at Darchula in western Nepal. Based on this earlier cooperation, the Norwegian company agreed to partner with NHE to jointly build the much larger (4 MW) turbines in Butwal.

During the project more than twenty NHE employees went to Norway for training with Sorumsand/Kvaerner for a period ranging from one to six months—including the chief design engineer who helped design the turbines and three welding technicians who received training and certification in the testing and quality control of welding. Manufacturing turbine casings requires highly skilled welders and machinists and strict quality control measures in which welds have to be examined (using X-ray and ultrasonic testing) centimeter by centimeter to guarantee that the components can bear up to enormous pressures. Aside from the turbine runner blades and some electronic control systems, all of the Jhimruk turbines were manufactured and assembled in NHE's Butwal workshops by Nepali technicians. This included the technically challenging welded steel spiral casings that had to be fabricated according to strict design specifications. At key points in the process Sorumsand/Kvaerner sent specialists to Nepal for short visits to supervise crucial steps. But generally it made more sense to send Nepalis to Norway for on-the-job training than to bring in foreign experts whose hourly rates could equal a week or two of a Nepali engineer's salary! The same was true of NHE's relationship with ABB Energi, the Norwegian generator and transformer manufacturer. NHE and BPC electrical engineers spent three-month training periods in Norway learning how to operate and maintain these electrical systems. Looking back at the extremely high-skilled technical experience that NHE gained on the Jhimruk project, Hoftun commented: "To learn those kinds of skills and be able to do that sort of job—that is the biggest, most important skill they have in NHE when it comes to power plants. No one else in Nepal comes close."[6]

The project's final challenge was to transport the finished turbines, generators, and other electrical equipment to the site, connect them to the penstock, set up the electronic control systems, and commence operations. Much of the installation was relatively easy. The exception was handling the power plant's massive, 32-ton transformer. This had to be delivered by road—a road that the government of Nepal had originally promised to build but by early 1994 still had not completed. When they finally finished the road in June 1994 and the transformer was successfully delivered to the site, it was only a few weeks before the target commissioning date of July 15, 1994. (The road washed away in the monsoon a few months later.) Still, under the close supervision of Kvaerner and ABB "commissioning engineers," BPC released water into the penstock for the first time in June 1994 and began live testing of turbines and power lines the following month. The plant went fully online on August 17, 1994, only a month after the planned date. At this point BPC management was extremely pleased with the project's progress: with Himal Hydro and NHE they had completed a complicated project with a high degree of quality and professionalism, basically on time, and well within budget.

According to its initial agreement with the Nepal government, BPC planned to operate the power plant for a one-year "guarantee period" after

---

[6] Hoftun went on to note proudly how the German contractors who undertook the 70 MW Middle Marsyangdi power plant hired NHE to manufacture the project's huge, bifurcated turbine inlet tubes and precision turbine casings. "This is where quality welding is crucial," said Hoftun. "But the Germans became so convinced of the quality of work done by NHE that kept bringing them more and more jobs to do."

commissioning so as to work out any kinks before handing it over to the government. During this first year, income from power generation would go to NEA with BPC refunded for operating expenses. Things started out well in August 1994 as monsoon rains powered the plant at full capacity for several months. But by October, technicians started noticing problems—to which I will return later.

RESISTANCE AND MITIGATION IN A "POLITICAL SPACE"

As any Nepali knows, the period of the Jhimruk project's gestation and birth—roughly the mid-1980s to mid-1990s—corresponds with one of the most important periods of social and political upheaval in modern Nepal's history. The Butwal Power Company's (BPC's) earlier projects at Tinau and Andhi Khola had benefitted from the tacit patronage of the Nepal government (which was officially a part-owner of BPC with a Nepal Electricity Authority [NEA] official chairing the BPC board of directors). But the government that approved Jhimruk in the late 1980s was very different from the government in place at the project's commissioning in 1994. Between those dates the massive Jan Andolan, or People's Movement, had toppled Nepal's thirty-year-old, unpopular, single-party Panchayat Democracy, restricted the power of the king, and installed a new multiparty democratic state that was still shaky on its feet.

Although you could hardly argue that Tinau or Andhi Khola had benefitted in any active way from the Panchayat state's autocratic and intimidating ways, Jhimruk was a slightly different story. As a state-commissioned project, at Jhimruk BPC felt that it was not their responsibility to worry about what local people thought of the enterprise. As with the rationale of all things the Panchayat state undertook, any suffering that locals might endure could be justified by the "greater good" supplied to the nation as a whole—in this case, mass power generation. The state had ways of dealing with people who disagreed with its decisions. But under the new post-Panchayat dispensation, all that changed. Local people were now empowered to stand up to the state and its unpopular initiatives—at Jhimruk, they did.

And they did so with good reason. Although it seems amazing in retrospect, construction at Jhimruk began without anyone having conducted even a rudimentary environmental impact assessment (EIA)—or study of the socio-environmental consequences of the project. In most parts of the world EIAs have long been required by law but in pre-1990 Nepal, this was not the case. It was only in 1991, after construction had begun, that the project's main funder, Norad—worried about negative consequences in the future—commissioned a Nepali consulting firm to carry out an EIA at Jhimruk. Only then did people fully realize the devastating effect the power plant would have on farmers downstream that depended on the Jhimruk Khola for irrigation water, especially during the dry season.

By diverting it under a mountain, through turbines, and into the Madhi Khola, the Jhimruk project removed significant amounts of water from a 25-kilometer stretch of the Jhimruk Khola above its natural confluence with the Madhi. In that 25-kilometer stretch, and especially in the 13 kilometers between the intake structure and the first significant downstream tributary, farmers depended on Jhimruk water, especially in the spring dry season when they used irrigation to produce a pre-monsoon rice crop to supplement a second monsoon rice crop and a winter dry-season wheat crop. In that 13-kilometer stretch alone farmers in twenty-three villages irrigated 164 hectares of land with dry-season Jhimruk water. Nevertheless, the Jhimruk power plant's original design intended to extract the river's *entire* dry-season flow for power generation. (In a land largely dependent on hydropower, the dry season is always a time of relative power shortage and therefore dry-season power production is the most valuable.) Ironically, while BPC's Andhi Khola power project (with its irrigation component) had allowed local farmers to *add* a third crop, the Jhimruk project would take away a third crop. As such, the project posed a serious threat to the livelihoods of thousands of local families.

Like coal mining, logging, commercial fisheries, or many other examples, hydropower production is a form of resource extraction—in this case removing water from a stretch of river in order to extract value from it. Furthermore, extracting a resource is an inherently political act in that the process of extraction does violence to existing local socio-environmental systems to the benefit of some (usually outsiders) and detriment of others (usually locals). For a resource to be extracted, the links between local people (living with the resource), the resource itself, and the local environmental and sociocultural systems that tie people to that resource have to be broken (or at least become strained). Some*thing* (ecosystems) and/or some*one* (people living within locally established socioeconomic systems) is likely to pay the price for the value extracted—in this case, hydroelectric power.

As such, hydropower projects almost inevitably produce what Shyam Bahadur Kunwar, in a fascinating paper analyzing the history of local resistance to the Jhimruk project (Kunwar 2016), calls "political space." The resource extraction zone becomes a political zone to the extent that the costs and benefits of the extracted resource—here water and power—are not fairly distributed among the "extractors" (here the government and BPC) and the "extractees" (local people). For that reason, except under the most authoritarian political circumstances, resource extraction projects almost always include some form of "mitigation"—efforts on the part of the extractors to compensate the extractees for the violence done to their lives and the value of the resource extracted. At BPC's Tinau site no real mitigation had been necessary because the entire power plant was in a rocky gorge devoid of agriculture and the whole project had been basically a win-win situation for BPC and local consumers. At Andhi Khola extensive mitigation features were built right into the overall project, with the power plant just one part of a large "integrated rural development" scheme that included irrigation, health care, education, and active local participation that successfully mitigated the impact of the resource extraction. Indeed if Andhi Khola is held up as an exemplary hydropower development success story, it is because the project largely succeeded in depoliticizing the politics of the extraction zone. By contrast, Jhimruk became a political space like few others.

Most people associated with BPC, including Odd Hoftun himself, tend to blame the Nepal government for Jhimruk's mitigation failures. For example, in a document dated February 14, 2001 (by which time Jhimruk had become something of a liability for BPC), Hoftun writes disapprovingly that

> the Jhimruk project has not been part of a district development program [as had been the case at Andhi Khola]. It was built by BPC as a turnkey contract job for HMGN [the Nepal government], starting from a pre-feasibility report, which turned out to have many flaws in it. The effects

of this are still felt both in the poor performance of the project, and in bad relationships with the local community.[7]

While it is easy to blame the government for its failure to carry out an EIA, it is also the case that Hoftun and other BPC leaders should have known better (and probably, at some level, *did* know better). Even though the government commissioned the project on the basis of a "flawed" pre-feasibility report, it is also true that BPC's Hydroconsult prepared the project's detailed technical plans before construction began, giving people in BPC plenty of time to contemplate the project's potentially disastrous local implications. And, in fact, BPC did add some (hopefully) mitigating elements to the project—most notably a relatively costly rural electrification component.[8] However, given that even the rather half-hearted mitigation efforts that were put in place came *after* the spring 1990 revolution, the timing suggests that BPC at some point realized that—under the new political system—it could not pretend that the project's violence was the government's fault alone.

At least some United Mission to Nepal (UMN) staffers working with the project sensed problems. Minutes from UMN meetings during the summer of 1990 record comments from UMNers critical of how the Jhimruk project was being carried out. One person calls Jhimruk "a lost opportunity" in that it focused only on "infrastructure development" and not potentially mitigating activities such as "community and individual development" as had been the case at Andhi Khola. Another commented that "Jhimruk was doing nothing for the local community and that, living in a separate compound, [we are] seen as intruders."[9] In fact, there are stark contrasts between the preparatory

---

[7] These lines are taken from pages 2–3 of a document entitled "Business Plan for BPC" which was included as Appendix VII in the bid documents submitted as part of the third round of the BPC privatization process—see Chapter 7.

[8] Yet, what BPC understood as an effort to share benefits (electrical power) with the community, local people just as often experienced as an effort to further exploit them by turning them into consumers.

[9] Hoftun archival material at Martin Chautari, folder: UMN Economic Development Board, September '83–March '87, Engineering and Industrial Development Department, Long Term Strategy—Kathmandu Discussion Group, notes of a meeting, June 29, 1990.

phases at Andhi Khola and Jhimruk. At Andhi Khola UMN or BPC staff were in the community literally years before construction began—living in local homes, establishing relationships, and mobilizing local participation and support. At Jhimruk, project staff (both Nepali and expat) lived in special project housing and it was only after construction had begun (and after the Jan Andolan) that BPC commissioned a drama troupe to perform a play intended to at least inform local people what was happening to them. In conversations with me Odd Hoftun acknowledged that some within UMN were "very upset" with the Jhimruk project by which, they felt, local people were being "cheated." Some UMNers even saw Jhimruk as a failure to act as Christians.[10] But Hoftun concluded our conversation by again stressing that

> Jhimruk was a *government project* and we were not able to act or change a lot in that original concept because the whole agreement structure was like that. BPC was just a contractor and we took it on in order that the companies should develop in contracting business as separate entities.

Ultimately one senses that Hoftun's priorities trumped any premonitions he or others might have had about the project's downsides.

Jhimruk's political implications would have been contentious under the best of circumstances (or best for BPC) but in the context of the political upheaval surrounding the People's Movement, the Jhimruk project became a lightning rod for political grievances. Coincidentally, Pyuthan District had long been a hotbed of political discontent and what had remained underground during the Panchayat years leapt into the open in 1990 when local labor strikes and political agitation delayed work on a key bridge being built for the Jhimruk project, leaving it unfinished before the onset of the monsoon. Several BPC staffers who had worked on both projects told me that at Jhimruk "the locals were not as cooperative as the people in Andhi Khola."

---

[10] "To be a Christian is to be a servant but are we perceived as servants by Nepali people [at Jhimruk]?" Hoftun archival material at Martin Chautari, folder: UMN Economic Development Board, September '83–March '87, Jhimruk Project Discussion, Engineering and Industrial Development Department, Long Term Planning, minutes of a meeting, July 11, 1990.

Some attributed this to the fact that, whereas at Andhi Khola the local people were largely *janjati*s (ethnic minorities, mainly Magar), at Jhimruk they were mainly Bahun/Chhetri—high-caste people more accustomed to advancing their interests and, if possible, working systems to their advantage. A more assertive population may partly explain the difference in local reception between Andhi Khola and Jhimruk but most important was the change in political climate: revolutionary fervor at the national level empowered local people to claim and defend their rights. And once parliamentary elections were thrown into the mix, political candidates quickly latched onto any and all local grievances in the hope of using them to their electoral advantage (Kunwar 2016). Whereas before 1990 local people had seen the Jhimruk project as a symbol of government indifference and oppression, *after* 1990 those criticisms shifted to BPC with the power plant emerging as a rallying point and target for local political activism.

With trouble brewing literally outside their gates, and with a full impact assessment finally in hand, in 1991 BPC had little choice but to shift tactics, trying to negotiate with local farmers and beef up Jhimruk's mitigation measures. Even though Jhimruk was still officially a Nepal government project, Kathmandu was now in such a state of turmoil that officials simply left it up to BPC to deal with local grievances. This both made BPC appear to be the source of all problems and effectively tied BPC's hands by leaving it with very little bargaining power in the absence of active state support. In hopes of quelling local resistance BPC offered to supply piped drinking water systems to fourteen of the most affected villages downstream. They agreed to maintain at least some year-round flow in the Jhimruk Khola to preserve aquatic life and offered to build a fish ladder on the intake dam. BPC hired a full-time agricultural specialist to demonstrate alternative farming techniques and new crops that would not require dry-season water. They built a special transmission line downstream to provide power to the most affected areas.

Of course a hydropower plant's most crucial resource is its water and, at Jhimruk, that too inevitably became a realm of mitigating concessions. A 2001 report on BPC's business prospects notes that, in the course of contentious negotiations both before and after Jhimruk's commissioning, local farmers and their advocates (politicians and NGOs) used "threats and pressure ... to dictate the terms of settlement ... which entails the complete shutdown of the plant for about one month each year, and that during the dry season when the value of the energy is at its highest." The same document acknowledges that these water concessions were about more than just irrigation: they were also due to a "lack of cooperation from the farmers' side and the deeply rooted hostile attitudes which have grown up." Repeatedly since 1994 local activists have "imposed their will on BPC by mobilizing crowds of people who ... physically force the power project staff to close down power generation if their demands are not fulfilled."[11] As much as claiming their rights to water, local people seemed to want to punish BPC.

Finally, BPC was forced to reopen the Pandora's box of issues surrounding land acquisition. Before 1990 BPC had left it up to a government representative, working with the local Panchayat chief district officer, to purchase land that the project would occupy (much like what had been done at Andhi Khola). Inevitably, after 1990 people came forward claiming they had been forced to sell land for unfairly low prices. Soon local politicians were pressing the issues, competing (for votes) to show who was toughest on the evil BPC and the foreigners who were trying to exploit them.

Discussions around land and other mitigation measures set in motion a process—nicely documented by Shyam Bahadur Kunwar (2016)—whereby negotiations between BPC and local stakeholders produced concessions that were followed by new demands, new negotiations, and new concessions. Eventually this pattern became institutionalized in a continuous dialectic between resistance and negotiation: after every negotiated agreement, politicians, NGOs, and other activists refocused resistance on a new grievance against the power plant as a way of perpetuating their political agendas. Through the 1990s, as Pyuthan District emerged as one of the principle Maoist staging grounds, the Jhimruk power plant became a favorite whipping boy for leftist NGOs and politicians. (And, as we will see in Chapter 7, a bombing target as well.) During the People's War local activists formed a permanent "struggle committee" to hold BPC's feet to the fire and to shut down the Jhimruk power plant whenever they did not get their way.

---

[11] From pages 2–3 of a document entitled "Business Plan for BPC" which was included as Appendix VII in the bid documents submitted as part of the third round of the BPC privatization process—see Chapter 7.

(According to Kunwar, this committee was active as recently as 2013.) In the past decade new NGOs promoting human rights in the area have used Jhimruk as an example of commercial exploitation by foreign interests, prompting still further rounds of resistance agitation. Having initiated the project with little concern for local sentiment, BPC seems to have parked Jhimruk under a permanent black cloud of resistance. Without the necessary mitigating measures, the inherent violence of resource extraction has boomeranged back to afflict BPC itself.

At the time of Jhimruk's commissioning in 1994, project accountants announced a roughly USD 1 million surplus remaining from the USD 19 million Norad grant. With an eye to further mitigation, BPC (along with UMN) used half of that money (augmented by a further Norad grant) to establish a small industrial development center—a sort of miniature version of the Butwal Technical Institute (BTI)—occupying the project's by then empty residential structures (Bakkevig, Hoftun, and Stensby 1996: 54). The Jhimruk Industrial Development Centre Pvt. Ltd. (JIDC) remains a going concern providing SLC-passed students with basic mechanical and electrical engineering skills. The center promotes "industrial and general rural and community development activities" as part of BPC's "continuing efforts to establish good relations with the surrounding community as well as the expanded use of electricity in the area."[12] On its website today BPC (ironically) claims Jhimruk Industrial Development Center (JIDC)—of which it holds 40 percent of the shares—as part of its "corporate social responsibility initiative." Perhaps more accurately JIDC is part of BPC's ongoing corporate self-inflicted disaster management initiative.

SEDIMENT

Another disastrous consequence of inadequate pre-construction technical assessment at Jhimruk concerns sediment. Himalayan rivers almost by

---

[12] From page 7 of a document entitled "Business Plan for BPC" which was included as Appendix VII in the bid documents submitted as part of the third round of the BPC privatization process—see Chapter 7.

definition carry heavy silt loads, especially during the monsoon when erosion is high. Based on past experience, engineers had designed the Jhimruk plant to remove sediment that can damage turbines. Two large settling tanks (36 by 5.5 meters) successfully removed 90 percent of silt particles larger than 0.2 millimeters. Normally these measures would have been sufficient to protect equipment from excessive abrasive wear, but it turned out that water in the Jhimruk Khola was not like that elsewhere. Subsequent analysis found that 80 percent of the tiny (0.2 millimeter or smaller) particles still suspended in the water when it entered the turbines were of extremely hard minerals like quartz and feldspar that easily abrade metals (Pradhan 2004: 4).[13] One Nepali engineer working at Jhimruk recalled, "No one foresaw the problems with sediment and erosion of the runners, not even the Norwegians. No one had tested the quality of the water in terms of sediment." Given that the problem was unprecedented, one can acknowledge both the Nepal Electricity Authority's (NEA's) and BPC Hydroconsult's failure to carefully analyze the Jhimruk Khola's sediment load as an honest mistake.

As noted earlier, after a successful commissioning in July and several months of strong monsoon power generation, by October 1994 Jhimruk technicians began noticing problems. Turbine bearings showed temperature readings that were too high and some of the sealing rings began to leak. Then, during the 1994 Dasain and Tihar holidays, when inexperienced technicians were left to manage as best they could, more serious damage occurred. When engineers shut down the plant for inspection they found faulty bearing temperature sensors and an inadequate bearing oil cooling system. But far more disturbing was the condition of the turbines themselves. After only a few months of operation they found that silt had caused "very severe" wear on many of the internal turbine components. In some places abrasive particles moving at high speed and pressure had cut grooves right through the 3-millimeter stainless steel inner turbine covers and guide vanes. On the high-tech runner blades themselves there was up to 1.5 millimeter

---

[13] On the Mohs scale of mineral hardness (in which 1 is talc and 10 is diamond), anything over 5 abrades metal, and quartz is rated at 7. Quartz particles as small as 0.006 millimeters (60 microns) can damage metal.

of wear with an average of 0.5 millimeter (Bakkevig, Hoftun, and Stensby 1996: A9-3). Even leaving aside its threat to the basic mechanical viability of the project's equipment, this turbine damage significantly decreased the Jhimruk plant's overall generating efficiency (Pradhan 2004; Bishwakarma 2007).

One by one the turbines had to be disassembled, taken to the Nepal Hydro and Electric (NHE) workshops in Butwal, and carefully reconditioned by building the worn metal surfaces back up to their proper specifications. Fortunately this exacting work could be done professionally *in Nepal* because having to send these parts to Europe for refurbishing would have been so expensive as to threaten the very commercial sustainability of the entire Jhimruk power plant. In rivers with abrasive sediment, conventional engineering wisdom favors Pelton over Francis turbines.[14] Both are equally susceptible to metal erosion but whereas technicians can exchange worn runner parts in a Pelton unit in about four days, to do the same for a Francis unit takes about four *weeks* (Bishwakarma 2007). But Jhimruk's 200-meter head means that Francis turbines are so much more efficient that their benefits outweigh their drawbacks compared to the Pelton design. All of this points to the fact that, in rivers like the Jhimruk Khola, engineers have to weigh the cost–benefit ratios of investing more money in sediment removal structures or in turbine maintenance (Bishwakarma 2007). Because none of these factors were known beforehand, the Jhimruk project is stuck with costly, complicated annual maintenance. Ever since its 1994 commissioning, the Jhimruk plant has been shut down for at least one month per year while its key components are rebuilt (Dahl 2014). But even so, the damage is so severe that the runner blades have to be replaced, at very high cost, every six years.

The silver lining of Jhimruk's cloud of problems may be that the plant has become a popular place internationally for sediment research. Nepali and foreign engineers associated with Hydro Lab in Kathmandu have amassed

---

[14] Pelton turbines are high-tech versions of the old-fashioned waterwheel—with concave "buckets" mounted on a wheel to catch water and turn a shaft. By contrast Francis turbines operate on a screw or propeller principle, like jet engines except designed to capture motive energy rather than generate thrust.

what one staff member referred to as "tons of data" on hydraulic flow, generating efficiency, and equipment wear as it relates to sediment loads. GE Hydro and Kvaerner use Jhimruk turbines to test proprietary coatings applied to runners to try to prevent erosion damage (so far with no real luck). Over time researchers have determined ways of maximizing overall efficiency at Jhimruk, even if it means operating at levels far below the plant's 12 MW generating capacity. According to a current Butwal Power Company (BPC) engineer, during the monsoon (when both power generation and damage potential are greatest) the plant runs at only 8 MW output and even that only under certain conditions. Earlier they would shut the plant down when water tests showed suspended particles of greater than 5,000 parts/million (ppm). "But now we are down to 3,000. The moment we reach 3,000 ppm, we reduce the generation because, above that, the erosion on the metal blades is so high that you are not getting benefit." Operating the Jhimruk plant now requires a careful balancing of the benefits of generating and selling power against the costs of damaging and repairing turbines.

JHIMRUK AND THE NATIONALIZATION OF BPC

When the government of Nepal commissioned the Butwal Power Company (BPC) to build the Jhimruk project in the mid- to late 1980s, the agreement was that, upon completion, the government would take over the plant. Paid for by a Norad grant, the plant was a gift from Norway to Nepal. After a one-year "guarantee period" BPC would walk away and Jhimruk would become one of several hydel plants (including Tinau) already government owned and operated.

But by the mid-1990s a number of significantly changed circumstances complicated the process of Jhimruk's handover. After the 1990 People's Movement the government itself was dramatically different, with power moving from one short-lived government to another as different parties vied for control. On the international development front, following the fall of the Berlin Wall and the collapse of the Soviet Union, western capitalist ideologues called for aid-based development (a system largely put in place during the Cold War) to be replaced by market-based development. In Nepal (as in many other developing countries) this "Washington Consensus"[15] increasingly made foreign aid contingent on various neoliberal "fiscal reforms" (Rankin 2004), including strong pressure to privatize state-owned enterprises. Foreign governments began pressuring Nepal to privatize entities such as the Nepal Electricity Authority (NEA) and its collection of hydel plants. Finally, in addition to a new national and international political environment, another factor complicating the Jhimruk handover by the mid-1990s was the project's by then all-too-apparent problems of bad community relations and sediment damage. For the Nepal government, the prospect of taking over Jhimruk, along with all its problems, was less and less appealing. Why take over these headaches if it was not necessary?

In the meantime Odd Hoftun, Balaram Pradhan, and others in BPC were themselves having serious second thoughts about the wisdom of handing Jhimruk over to the government. Having watched the sad decline in conditions at Tinau, BPC leaders were not eager to see the same government negligence bring down Jhimruk as well. Another former BPC engineer I interviewed estimated that, at that time, NEA suffered transmission power losses of about 45 percent because of poor maintenance and management. With all these factors in mind, in July 1993 BPC itself approached the government with a proposal that, rather than take over the plant, it could leave Jhimruk in the hands of BPC as owner/operators and instead take shares in the Jhimruk project. BPC also offered to privately manage transmission and distribution in a four-district area surrounding the plant.

Negotiations proceeded until—unfortunately for BPC—just weeks before the scheduled handover in July 1995, the then minister of water resources announced that the government intended to take over not only the Jhimruk power plant but the entire BPC as well. BPC's major shareholder, the United Mission to Nepal (UMN), would hand over its shares to the government which would then maintain BPC as a quasi-independent "public company" under government ownership. Alongside NEA, BPC would continue to own power plants (including Jhimruk) and operate as a (formally) independent

---

[15] According to which "policy actors, even if they are not themselves market actors, agree that market-like mechanisms ... are what are needed" (Thompson, Gyawali, and Verweij 2017: 11).

power producer and distributor. Under this plan UMN would retain less than 3 percent of BPC shares, although it would be allowed one basically ceremonial seat on the new government-controlled BPC board.

At least from within the BPC fold, "this was an outcome which all concerned had agreed should be avoided by all means" (Bakkevig, Hoftun, and Stensby 1996: 53). Losing control of Jhimruk was unfortunate but losing control of BPC was disastrous. From the beginning Hoftun had known that the key to BPC's success would be to maintain firm corporate control in order to uphold the company's mission of incremental growth, reinvested profits, and ethical business practices. Also from the beginning the Nepal government had been a part owner of BPC with the company's board chair a government appointee. But Hoftun knew that if and when the chair's opinions differed from those of the BPC management, the government's position would be overridden. Government involvement: good. Government control: bad. But even so, in interviews with me, Hoftun admitted that the government's move made sense from their perspective. Trying to manage Jhimruk on their own would have been challenging. But by maintaining BPC as a legally private corporate entity (rather than absorbing its assets into NEA), the government (as majority shareholders) could both reap the benefits of controlling BPC's profit-earning potential *and* absolve itself of the responsibility of managing the challenging Jhimruk project. The government could watch BPC do the work and then skim off the profits.

Due to a string of disappointments and personal tragedies (discussed in the following chapter), by the time final negotiations between the government and UMN were being hammered out in 1995 and 1996, Odd Hoftun had retired (more or less) and returned to Norway. Hoftun tried to make his opinions known but in his absence, and as the principle BPC shareholders, it fell to UMN representatives to determine not *if* but at least *how* BPC would be transferred. From a legal standpoint UMN had no choice but to comply. Eventual government ownership had been written into BPC's charter from the day it was incorporated in 1965. But the *nature* of that ownership was not at all predetermined. Might there be ways of both satisfying the government *and* maintaining independent board control of BPC?

Odd Hoftun and Balaram Pradhan certainly thought so. Yes, the current minister might be applying pressure on UMN to make the transfer straight up. But there were ways of making negotiations proceed slowly and, with governments changing at short intervals, it was not unlikely that in a few months or a year a more amenable minister might be in place. After lifetimes of engaging state officials, both men knew how the government worked and both had friends in high places such that, with enough time and finesse, they knew that acceptable and honorable agreements might be arrived at. By contrast, the person appointed by UMN to negotiate with the government was new to Nepal and eager to comply with government requests.[16] As Hoftun delicately explained it, the UMN negotiator

> had a western respect for government decisions which we who had worked in Nepal were not so, um…. Well, we knew that things could be changed. By having good contacts and talking with people, one could have … I mean we had done that in a couple of cases over the years…. [But the UMNers] were much too eager to comply immediately, get it done. [They argued] "That's how it is and the law and agreement is clear: when government asks, then we should hand it over."

When I asked Hoftun what he would have proposed instead, he said simply, "Taking time. A year afterward there would have been new people in the government … and we could have negotiated a better deal." As an example he mentioned how, in response to government pressure to nationalize the Butwal Technical Institute (BTI), BTI officials had managed to delay and eventually negotiate a deal whereby the institute became officially state owned but retained an independent board made up of representatives of relevant interested parties. Such an arrangement in which an independent board had "full freedom to run the project although the government remained the formal owner" would have been, for Hoftun, an acceptable fate for BPC.

When hopes for keeping board control over BPC seemed doomed, Hoftun lobbied hard for another means of keeping BPC out of government hands, this one involving selling BPC to private investors and giving the cash

---

[16] Hoftun also suspects that at least some within UMN were eager to wash their hands of what was, to them, a (by definition) morally suspect commercial operation like BPC.

proceeds to the government. Given the current international development climate that was pressuring Nepal to *privatize* state-owned businesses, the government's move to *nationalize* BPC was suspect and went against even the Nepal government's own officially stated position. The minister of water resources was promising to soon privatize BPC but Hoftun and others I interviewed suspected that, once the company was in government hands, the minister had more personal financial transactions in mind: if he could control the privatization process it would mean substantial bribes and kickbacks for himself and his cronies. But if UMN, through an impartial third party, could sell BPC in an open bidding process then the proceeds would go to the government, not the minister. Hoftun believes that if BPC had been sold to private investors in 1995 it probably would have brought a higher price than when the government finally did privatize BPC almost a decade later (see Chapter 7).

But UMN disregarded Hoftun's and others' pleas and on August 16, 1996, completed the official transfer of BPC shares to government control. Hoftun was bitterly disappointed. Dan Jantzen, a close collaborator, saw the decision as ill considered, precipitous, and fateful. Speaking for Hoftun as well, he said, "In my mind, that single decision changed the subsequent course and direction of BPC.... How much better it could have been if the takeover was done gradually and in a controlled manner.... [T]here is no reason why it had to be this way" (Svalheim 2015: 273). After nurturing the company for thirty years, Hoftun, Pradhan, and other like-minded BPC leaders had lost control of BPC and its mission. They then spent the next two decades struggling to regain control of BPC, or at least exert some degree of influence on the company's corporate culture.

Once in government hands, conditions at BPC deteriorated rapidly. Not wishing to work for the government, Balaram Pradhan resigned (though he did serve as a government advisor and NEA board member during a brief Unified Marxist–Leninist (UML) Party administration before a stroke forced him out of public life and into a long recovery). Other Nepali engineers soon followed due to a lack of work and bleak career prospects. One person described a "slack period" after the government transfer when "new projects did not come. The government representatives on the board could not keep BPC a priority in their minds or in their programs. So BPC did not have any new activities for several years after the handover."[17] With only "a very small amount of work" to do, he quit. When I asked whether this stagnation was because of poor leadership or because, by the time the handover occurred, the government was embroiled in the Maoist People's War, or Civil War, he admitted that the Maoist conflict was certainly a factor. "But the primary reason was a lack of vision from the leadership. Because whoever was in leadership, they were from a government background and BPC was not a priority. [As BPC board members] they were just representing some government department or ministry" rather than really providing active leadership for BPC. For people appointed to the BPC board, service was at best often an appendage to their other more political concerns and at worst, a simple nuisance. (At least in part this was due to the structural redundancy of the government's owning two more or less parallel entities: BPC and NEA.) Having lost most of its old leadership, and without the government being able or willing to step in, it was as though BPC had its hands tied and could do nothing.

From the beginning, foreign engineers assigned through UMN had been an important resource for BPC and a key part of Hoftun's vision of technology transfer to Nepalis. As per its agreement with the government, and at Hoftun's strong encouragement, UMN continued to second expat engineers to BPC even after the takeover. But those foreigners who took BPC assignments found the environment less and less comfortable. BPC's Hydroconsult division retained an active foreign work force but within the main part of BPC, foreigners soon found themselves to be unwanted. As one expat former engineer recalled:

There was a fairly strong anti-foreign sentiment in BPC at that time. Particularly as these NEA bosses would come in [to see foreigners working] and say, "What's this nonsense?" There was a feeling that the time had come. "We're a Nepali company now. It's time to move on."

---

[17] At the time of the handover BPC was deeply involved in the Khimti project (described in the next chapter) and would remain so until the plant's commissioning in 2000. However, with no new projects to develop, BPC engineers were left with little to do after 1996.

Ok, fair enough. That was understandable. Those of us in BPC got the message loud and clear. So I didn't opt to come back to BPC after I went on furlough.

Odd Hoftun had hoped that continued UMN expat presence in BPC would help to keep his more service-oriented, humanitarian vision alive with the corporation, and he often personally encouraged people to stay with BPC—with increasingly little effect.

This chapter has traced the BPC story from the early 1980s and the initial planning for Jhimruk, through the project's completion and complications, to BPC's nationalization in 1996 and beyond. But during that same time BPC had launched its last Hoftun-initiated project. The 60 MW Khimti project would strain Hoftun's development philosophy to the breaking point and provide pivotal lessons for BPC and its collaborating companies Himal Hydro and Nepal Hydro and Electric (NHE).

5

THE "GREAT UPHEAVAL"

Khimti and the Limits of the Hoftun Hydropower Vision

> It is hard to imagine how any financial arrangements for hydropower projects could be more complicated than what was the case for the Khimti project, both with regard to the financial and the legal and regulatory sides of things.... In the end this amounted to more than seventy separate agreements which all had to be tied together in a complex package.
>
> —Odd Hoftun[1]

The 60-megawatt (MW) Khimti hydel project represents a fundamental turning point in the history of hydropower development in Nepal. Whereas almost all of Nepal's previous projects were government and/or grant funded, Khimti was the first to be developed by private investors and the first to involve extensive collaboration between a Nepali company—the Butwal Power Company (BPC)—and international commercial developers. In keeping with the then ascendant neoliberal development philosophy of "unleashing market forces," and with post-Andolan Nepali governments eager to comply with the wishes of international donors (Gyawali 2003: 77), in the early 1990s Nepal put in place a series of laws that laid the legal foundation on which private-sector commercial power projects could be built. The Khimti project precipitated this new legal

---

[1] From a document (headed 010207 Section 1 and 2-dj) entitled "Section 2. The Strategic Investor/Purchaser," part of the bidding documents, third bid, 2001.

context but in so doing opened up Nepal to the gradually building, and now flourishing, market of independent power producers (IPPs) that are today the leading force in Nepal's power sector. Because of BPC's push to develop Khimti, Nepal had established the legal framework for private-sector power development more than a decade before India and other Asian nations. And because of BPC's and its daughter companies accumulated expertise and established human capacity in project development, the stage was set for Nepali manpower to continue to independently develop Nepal's hydropower potential.

But the moment that saw the birth of Nepal's private hydropower development sector was also, in some ways, the end of Odd Hoftun's development vision. At 60 MW, the Khimti project was the logical next step in Hoftun's plans to incrementally grow Nepal's human capacity in the hydropower sector. But a 60 MW project also, for the first time in Hoftun's career, finally surpassed the point at which government or donor grants could finance his undertakings. At this scale, project development would require international commercial financing which, in turn, would put in place a very different set of corporate dynamics. With international players investing major sums of money into "his" project, Hoftun quickly found himself overshadowed by more powerful players and his development vision supplanted by corporate logics that no longer understood Nepal to be the prime beneficiary. Hoftun launched the Khimti project in the early 1990s but by the plant's commissioning in 2000, his presence had been reduced to that of a figurehead. Similarly BPC went from being the project's sole developer to owning only a 15 percent share in the final product.

One Nepali engineer who worked closely with Hoftun for decades likened him to a chess master who played the game from a superior vantage point. "Odd had a 30,000-feet perspective. He was here to help Nepal, to develop Nepal. BPC, NHE, and the others were his chess pieces." From his lofty viewpoint Hoftun also saw Nepali society, Nepali political systems, and even individual politicians as other pieces at play in the game. After four decades of work in Nepal, Hoftun knew that all of these pieces had to be strategically accounted for, outmaneuvered, and perhaps (subtly) played to one's advantage—in order to achieve the greater goal of the greater good.

But what Hoftun did not adequately foresee was that, with the arrival of international actors and corporate forces on the hydropower development scene, the chessboard suddenly grew to a global scale that he could no longer play. On that global board, Odd Hoftun, BPC, and Nepal itself all became pieces to be played by forces far beyond their control.

INCEPTION

By early 1990 the Butwal Power Company's (BPC's) Andhi Khola project was almost finished, Jhimruk was underway, and Nepal was in the throes of the first Jan Andolan, or People's Movement. Once the old Panchayat government had taken its place in the dustbin of history, the new Nepali Congress government was desperate to push forward on national priorities, with power generation near the top of the list. It was in that context that, in September 1990, Secretary of Water Resources B. K. Pradhan approached Odd Hoftun at a Jhimruk coordinating committee meeting and delivered something between a plea and a command to not only hurry up with Jhimruk but also find something "five times as big" and build it. "The country is suffering a power crisis. Hurry!" (Svalheim 2015: 257). This was an attractive proposal to Hoftun who was himself already looking for a new project—both as part of his larger vision and, much more practically, simply in order to keep an increasingly large numbers of skilled workers actually employed. If he could get the project up and running quickly, Himal Hydro crews could move straight from Jhimruk to Khimti, with Nepal Hydro and Electric (NHE) soon to follow.

Even before getting the secretary's directive, Hoftun had his eye on a site on the Khimti River east of Kathmandu (based on a pre-feasibility study prepared by the Nepal Electricity Authority [NEA]) and was discussing a potential project with BPC and United Mission to Nepal (UMN) colleagues. Although UMN support was weak (see following), with government encouragement, Hoftun proposed to develop the project—which would increase Nepal's overall energy supply by 25 percent—and was given immediate authorization. By December 1990 Hoftun was in Oslo negotiating with Norad, which then provided money for a detailed, technical, "bankable" feasibility study that BPC Hydroconsult carried out in early 1991 in cooperation with Norpower, a Norwegian hydel consulting firm.

As results of this technical study came in, it became clear to Hoftun that BPC alone could not handle Khimti's large-scale challenges. Khimti would indeed be five times larger than Jhimruk and Hoftun knew that, even with Jhimruk under their belts, BPC, Himal Hydro, and NHE did not have the expertise needed to pull off a project of that scale and complexity. Money was also a major concern. Hoftun hoped that by using Nepali manpower and the cost-saving methods he had honed during his earlier projects, Khimti could be built relatively cheaply. But at this scale even the most frugal project would require funding many times beyond Jhimruk's budget, beyond what anyone could hope to acquire through grant funding alone.

What's more, support from UMN—always half-hearted at best—was beginning to disappear altogether. Minutes from UMN Economic Development Board meetings in 1990 record voices critical of the prospect of UMN supporting Hoftun's vision beyond Jhimruk. One person pointed to a disconnect between UMN's stated mission to "lay stress on poor people" and the construction of major hydropower plants to "build national infrastructure," noting sarcastically that "either we should amend our objectives or we should tell the government that hydropower is not our calling and decline to take on a successor project to Jhimruk." Another questioned the appropriateness of tackling large-scale projects: "Up till now UMN has been meeting a national need, because international companies were not interested in coming to do small-scale work. But as we get bigger we are competing with internationals. Is this our aim?"[2] Ultimately UMN continued to support Hoftun by supplying skilled foreign missionary volunteers to BPC, Himal Hydro, and NHE but the Mission was no longer willing to fully engage with projects as they had done at Andhi Khola (and would have done at Jhimruk). In short, with Khimti Odd Hoftun found himself unable to count on his accustomed sources of support: the project

---

[2] Hoftun archival materials at Martin Chautari, folder: UMN Economic Development Board, September '83–March '87, Engineering and Industrial Development Department, Long Term Strategy—Kathmandu Discussion Group, notes of a meeting, April 7, 1990.

was too expensive to expect Norad or the Nepal government to fund and too morally ambiguous for UMN to embrace.

In September 1991 Hoftun was sitting in his Butwal BPC office trying to imagine a way forward for Khimti when a group of Norwegians arrived, unannounced, at his door: a delegation from Statkraft, the Norwegian state-owned hydel development corporation. By the 1980s Norway had developed virtually all of its water resources for hydroelectricity, finding itself with a mature and sophisticated hydropower development industry—with no more work to do. In response, the government organized its hydel development resources into a state-owned corporation named Statkraft that was authorized to operate commercially both in Norway and, for the first time, abroad. The delegation that knocked on Hoftun's door was part of a mission "to search the world for new markets for Statkraft's specialty: To design, build, and operate new hydropower projects" (Svalheim 2015: 247). With its huge hydropower potential Nepal was a perfect market for Statkraft's skills and services but, with no previous international experience, the company had no idea of how to establish a foothold.

At the moment of BPC's and Statkraft's meeting both sides had what the other desperately needed. For BPC, a partnership with Statkraft would provide exactly the kind of technical expertise needed to pull off the Khimti project. What's more, Hoftun figured, with its deep pockets and huge corporate assets, Statkraft should be able to invest its money in Khimti, or at least offer its assets for bank collateral. For Statkraft, a partnership with BPC provided a safe channel into a new, potentially lucrative market: a local company that would help pilot them through Nepal's legal and cultural shoals—and one headed by a Norwegian to boot! By late 1991 the BPC board had authorized Hoftun to negotiate a partnership with Statkraft and by early 1992 BPC and Statkraft had agreed to establish a joint-venture company—Himal Power Limited (HPL)—to build, own, and operate Khimti. Each company would own 50 percent of HPL's shares and Hoftun would be HPL's first general manager.

As BPC and Statkraft representatives traveled feverishly back and forth between Norway and Nepal trying to get the Khimti project moving in the first half of 1992, they faced challenges far beyond the technical and logistical issues of simply building a power plant. Foremost was the simple fact that the legal environment needed for private-sector international investment in hydropower production did not yet exist in Nepal. There was no legal framework regulating how a private commercial developer would use public water resources (such as the water in flowing rivers) and there were no laws spelling out how a private power producer would relate to the government's existing electricity production and distribution system. Because Norway had long experience with these matters, BPC and Statkraft brought in a Norwegian legal delegation to help the Nepal government draft legislation that would enable, but also regulate, international entities wishing to do hydropower business in Nepal.

Originally envisioned as the Khimti Act that would have determined conditions for a single project, Nepali officials quickly realized that the law should be drafted more broadly. As a result, in 1992 the government authorized a new broad Hydropower Development Policy as well as two new pieces of legislation: a Water Resources Act to regulate the fair distribution of benefits from the private use of public water resources and an Electricity Act laying out how NEA would relate to independent power producers (IPPs).

> The unveiling of these three documents was an important milestone in the history of Nepal's power sector because it heralded entrée of private investment in the power sector, thus a significant part of infrastructure sector. This is the first policy related to the power sector in Nepal, which sets out the modality that uses public resources and private investments to deliver services. (Shrestha 2016)

In an interview with me, Ratna Sansar Shrestha pointed to Khimti as "the milestone that got us here. So if Odd hadn't gotten this rolling by now we might have an electricity policy, but it would be a carbon copy of the 2003 Indian legislation. But because of Odd's push, we had it eleven years before India." This 1992 legislation opened the door not only for BPC and Statkraft but for the entire now-booming industry of independent power production in Nepal.[3]

---

[3] Also critical was the move instigated by the late Shailaja Acharya who, as deputy prime minister and water resources minister, in 1995 forced NEA to announce a buy-back rate for private hydropower producers, thereby encouraging private investment.

The year 1992 was also one of personal tragedy for Odd Hoftun and his family. On July 31 Odd and Tullis's second son, Martin—then twenty-eight years old and a doctoral candidate at Oxford writing a dissertation on politics in modern Nepal—was killed along with 113 others in the Thai Airlines crash near Kathmandu.[4] Coping with Martin's death could only have made the other setbacks that Hoftun confronted in the coming years more difficult to bear.

Even without financing in place, in late 1992 Hoftun authorized BPC to start preliminary work at Khimti. In light of Nepal's debilitating energy shortage, Hoftun felt it was in the national interest to push the project forward. But he was also motivated by the fact that Himal Hydro crews were winding down their construction work at Jhimruk and needed new assignments. Drawing on money coming in from the by-then operational Andhi Khola power plant, BPC was able to hire Himal Hydro to begin developing the project site. This involved beginning construction on the powerhouse access tunnel, the spoils from which were used to raise and level a large area of previously more or less useless gravely wasteland near the village of Kirne. On this new land Himal Hydro began building workers' quarters and administrative buildings. By launching the Khimti project using BPC's own limited funds before project funding was in hand, Hoftun showed that he was willing to shoulder risks for the benefit of Nepal, rather than the benefit of business or investors. At that point Hoftun did not realize how controversial this would soon be.

What no one realized at the time was how the same post–Cold War political and economic currents that were sweeping the world—triggering revolutions from Eastern Europe to Nepal and putting in place a new neoliberal aid and development policy—would also transform Statkraft itself. Hoftun felt this viscerally in what he experienced as an almost total change in corporate culture within Statkraft in the years immediately following their joining of forces in HPL in 1991. Under the then ascending global doctrine of neoliberal restructuring (that called for, among other

---

[4] In Kathmandu Martin Hoftun's legacy lives on in Martin Chautari, a now prominent public research institution and discussion forum that Martin, along with Nepali and foreign colleagues, had founded in Kathmandu in the 1980s.

things, the privatization of national assets), legislation in Norway opened up the domestic power market to private competition and required Statkraft to spin off its construction and engineering divisions into separate commercial entities: Statkraft Anlegg and Statkraft Engineering (not unlike how Himal Hydro and BPC Hydroconsult were the offspring of BPC). But more to the point, what had been state-owned entities focused on overcoming engineering challenges now became privately owned commercial entities focused on profitability.

Hoftun experienced this change as a traumatic shift from dealing with fellow engineers to dealing with corporate lawyers.

> When we started talking with Statkraft—that was before the Great Upheaval!—the people were old-timers who were thinking in the same language I was used to and the same concepts. You know I grew up in a hydropower family, so we were used to ... I mean, there was an element of trust between contractors and owners. So you didn't make elaborate contracts with lawyers and all that. You just agreed on something and you stuck to what you had agreed to.

He went on to explain that there were, of course, contracts in the past but not wildly elaborate ones:

> There wouldn't have been lawyers involved except when there was disagreement about interpreting something written down. There was a written contract but it was between people thinking like engineers, not like lawyers! So that was a big change and the other big change was commercialized business—new visions and principles—which in turn resulted in a complete change-over at the higher level of management at Statkraft. They just went in and discharged all the older people and got in new ones from the new business world.

"So literally the people you were dealing with at Statkraft changed?" I asked. "They changed. All of them practically, at that level. Earlier it was engineers who were in charge of projects like this. But [since then] the finance people have taken the lead in thinking." Hoftun no longer spoke the same language as most of the people who now served on the HPL board.

When BPC and Statkraft agreed to form HPL they had a shared vision of the technical challenges involved but as Statkraft itself morphed into a new corporate entity major differences appeared with respect to how the Khimti project would be financed. Hoftun assumed that an enormous partner like Statkraft would be able to finance Khimti largely through its own equity investments and using its assets as loan collateral. With good government relations in Nepal, BPC could ensure that the project would proceed smoothly and at minimal cost.

But the new Statkraft had other ideas. For them, their first foray into international business was an opportunity to create ties with major development banks such as the World Bank's International Finance Corporation (IFC) and the Asian Development Bank (ADB). Rather than fund Khimti directly, Statkraft wanted to take out bank loans to establish a credit record for future projects. However, to their "great surprise," Statkraft learned that—as first-time international borrowers—they were not creditworthy, despite the fact that they were a major corporate entity with huge assets (Svalheim 2015: 262). Therefore, in order to protect their loans against the risk of lending money to Statkraft, the banks demanded enormous and costly *over-financing* to protect their money against every conceivable and inconceivable risk—thereby massively driving up the cost of the overall project.[5] From the original cost estimate of USD 90 million, the final amount rose by over 50 percent to USD 140 million. In short, the project's bank financing was so loaded down with insurance, protection, and

---

[5] There are various ways to insure a project against loss, one of which is to have one or more of the project's developers provide a "corporate guarantee." In the case of Khimti, Statkraft refused to put up their own cash resources as collateral for loans. Instead, in a system known as "Project Financing," the *project itself* was offered as collateral such that, if things fell apart, the banks could at least claim the project. But this is risky for the banks because what happens if the project is never completed? The collateral itself is not a sure thing; hence things have to be "over-financed."

contingency guarantees that it drove the cost of *funding* the project far, far beyond the actual cost of *building* the project.

Hoftun vigorously objected and pointed out that simple logic called for the project's risks to be borne by the people in whose interest it was to minimize risk—namely, the developers themselves. Hoftun had decades of experience following frugal and conscientious development practices in Nepal. He knew that a lower-cost project would result in lower-cost power, which would be good for both the plant's owners (HPL) and consumers in Nepal. Why build an artificially expensive power plant simply in order to meet unnecessarily inflated debt obligations? But ultimately it was the banks that called the shots. "In reality everything was decided not in the BPC board but by the banks," said Hoftun. Due to their inexperience with, and apprehensions about, investing in a place like Nepal, the banks insisted on layer upon layer of risk protection, thereby driving the price of the project and its product—electricity—to a price that is still controversial to this day. But Hoftun blames Statkraft: "It was Statkraft who insisted on having them [the banks] in. And that was for *their benefit*, for their own benefit."

In a taste of what was to come, Statkraft overrode Hoftun in the HPL board and, exercising their *"more* equal" status in their "partnership" with BPC, went ahead with the over-the-top bank financing plan (Svalheim 2015: 262, italics added). Simultaneously, in order to enhance HPL's status in the eyes of international lenders, Statkraft insisted that BPC's 50 percent interest in HPL be reduced to 30 percent. And as a final insult, HPL's creditors demanded that tiny BPC deposit a "completion guarantee" of over USD 2.2 million in an American bank as "contingent equity" to be drawn on in case BPC could not meet its loan obligations—thereby placing BPC, in Hoftun's words, "on the brink of insolvency." None of these moves served Nepal's interests but they did serve Statkraft's long-term corporate interests.

One of the first victims of BPC's new financial insecurity was the preliminary work being done at the Khimti project site—which came to a sudden halt. In fact, as the HPL board's guiding principles had shifted, Statkraft officials became increasingly unhappy that preliminary construction had ever even begun. "There were some *heated* [HPL] board meetings on that subject," Hoftun recalled.

> They thought we [BPC] were sabotaging their efforts to get the best possible deals with the government, to get the power sales agreement worked out. This advance work suggested that we were being too friendly to the government. We were on the *wrong side* of the conflict! Basically we were thinking like Nepalis.

In other words, by going ahead with the project before its financial implications were fully worked out, BPC was suggesting that getting the project built was more important than negotiating the highest profit margins for HPL. Worse still, Statkraft argued, preliminary construction actually undermined HPL's bargaining position by making it appear less likely that HPL would pull out if its conditions were not met.

Another victim of these Statkraft- and bank-induced woes was, eventually, Hoftun himself. As general manager of both BPC and HPL, and as negotiations with Statkraft became increasingly difficult, Hoftun finally arrived, as he put it, "at the point where I found it impossible to deal. I found it impossible to deal with Statkraft and they followed the bank." What's more, Hoftun felt drained from an emotionally and physically exhausting period: he had lost his son and he had lost control of the Khimti project, in spite of having made more than thirty trips between Europe, Nepal, and Manila trying to negotiate favorable terms in the last four years before his retirement. Looking back, Hoftun recalled how those involved in the Khimti negotiations "were unlucky enough to be doing this at the most volatile time. We had a front row seat! And I felt it was so difficult that I brought in someone else." That person was Peter Harwood, a retired British Shell Oil executive with extensive international business experience. "He was a good man," recalls Hoftun. "He knew how to bang the table and talk up against them in a manner that I couldn't.[6] And I was

---

[6] A retired NEA official I interviewed recalled Harwood as a determined, seasoned negotiator. "Harwood? I used to call him *Hard*wood! He was really, really tough."

happy to retire. My time was out." In July 1993 Harwood replaced Hoftun as BPC general manager and nine months later, in March 1994, took over from Hoftun as general manager of HPL, leaving Purna Prasad ("P. P.") Adhikari in charge of BPC. Harwood worked to defend the interests of BPC and its corporate offspring but, in the end, much of Hoftun's development philosophy was brutally pushed out of the equation by the interests of multinational capital.

The gradual takeover of HPL and the Khimti project by international financial interests was a bitter disappointment not only for Hoftun but also—and maybe even *more so*—for the rest of the BPC team. All of the Nepali former staff members I spoke with pointed to the decision to halt construction at Khimti (following BPC's bank-required "completion guarantee" cash deposit) and Hoftun's resignations from BPC and HPL as not just the end of an era but the end of a vision or mission that they had collectively embraced. Several people spoke of the halting of preliminary construction as the beginning of a "gap" after which everything had changed.

If it was hard for Hoftun to make sense of what was going on, it was even harder for the people at work on the project site. Some people feared for BPC's future after Hoftun's retirement. "Actually in BPC we were *very worried*," remembered one then junior Nepali engineer.

> Odd Hoftun had started BPC. He continued it for a long, long time, very smoothly, very successfully. We were worried that once he leaves, BPC may collapse. We were just young engineers. We didn't know if the system could survive without Hoftun.

Some felt a sense of betrayal by the Norwegians. Up to that point, Norwegian aid money had mainly footed the bills for BPC projects. Now the Norwegian Statkraft was involved but they were putting the entire project at risk. Norwegians had been partners but now seemed threatening and self-serving. As one Nepali engineer recalled:

> If the Norwegians had wanted to put money into it, then they would have done it already and the project would have gone smoothly. "Maybe they don't want to come," [people thought]. "Maybe they want to return back to

Norway and the project will never come." That's the kind of very negative feeling that was there. At that point there was very little hope that the project would be implemented.

Other Nepali staff members experienced the halting of construction at Khimti as the loss of a sense of membership in a common cause. As veterans of previous BPC projects, said one person, "We felt that we were working on a *team*. We were cooperating with each other—Nepali and foreigner, BPC and the other companies. We felt like we are a *family* so we had good teamwork." When the word came down to stop work at Khimti—with no promise of if or when it would resume—it was as though one's family was being broken apart or one's team was being disbanded.

This sense of gut-wrenching loss and disappointment comes through in the comments of one Nepali engineer. He describes the feelings of BPC and Himal Hydro veteran staff then working at the Khimti site.

> In a way, we were working on a *mission*. We were *developing*—5 [MW] to 12 to 60 to 100, 200, 500. So it was not like just building a single project. We were guided or trained in a way that we thought we would be growing continuously, and that our growth would be in an accelerated manner. That was our thinking. After completing Khimti we'll go to … some other project.

He continued,

> But when we had to stop Khimti it had a very bad impact on people working for the project. When we had to lay off people and tell the contractor that we had no money and so can't continue, then all those people and equipment had to be taken away. *Those people who were working, and feeling kind of ownership of the project.…* Well, suddenly, when they had to leave their job and come back to their home or to their company where there was not enough work—some people started to cry actually. It was a very emotional situation. They would go the project site, take photographs to have some record of it. Everybody had tears in their eyes because they were leaving the project. *They were on a mission and when you are losing the battle,*

you can imagine what happens. It was like that. It was that feeling. (Italics added)

As I suggested in Chapter 1, a big part of what made BPC successful, and a major reason why Nepalis committedly drove the institution forward, was this sense of "mission." This was not the "mission" of the United Mission to Nepal (UMN), but it was based in Odd Hoftun's development philosophy—a vision that closely complemented the goals and values of the many Nepalis who chose to join the BPC "family" or "team." Rather than thinking of themselves as "just building a single project," they saw their work as acquiring experience, continuously growing, and, crucially, serving the nation. There was something visionary, historic, and morally good about what they were doing. They were developing Nepal—bringing not only power but also water, increased food security, employment, skilled manpower, and so on. There was a sense of being a part of something much bigger than themselves. They were part of a "mission" that everyone could feel good about. Profit generation was clearly a part of this mission or vision, but there was the clear sense that the profit would be plowed back into the companies to produce "accelerated growth."

Several BPC veterans recognized this moment as the point at which this sense of mission dies, or at least begins to be extinguished. A project that began with the team spirit of working for the common good was taken over by forces driven by a very different mission: to make money for banks and other investors. Rather than being plowed back into the company like fertilizer, now profits were to be extracted—the more the better. Looking back, another Nepali former Himal Hydro engineer lamented:

It should have been like Odd Hoftun's vision [of growth and capacity building]. If that had continued on, with the same group of people and sister companies, it would have been so fantastic for Nepal. But with Khimti, that ended—for all these reasons—and then nobody knew what to do. The government lost the momentum that BPC had built up.... *You know earlier there was a vision, there was imagination. "This is how we should go," and I was trained in that way. But then there was this gap and suddenly the whole thing vanished.* (Italics added)

I think it is important to highlight the fact that this vision—shared by Hoftun and so many members of the BPC family or team—was shattered not by "corrupt Nepalis" or "Third World incompetence" but by international corporate capitalism and commercially driven "development" banks.

REACHING FINANCIAL CLOSURE

Of course construction did eventually begin again but only after literally years of legal wrangling—between Statkraft, the banks, Himal Power Limited (HPL), and the Nepal government—finally ended with "financial closure" on January 15, 1996. A process that Hoftun had expected to take one year in the end took four years to complete. Because the World Bank insisted on following American legal procedures, Statkraft hired a high-powered American lawyer to steer a course through the endless negotiations, legal minutiae, and paperwork required to nail down the project's finances. The lawyer's fee? USD 2 million.

In fact, because the banks did not have confidence in Nepal's legal system, the Khimti project required still another piece of official legislation in Nepal. Up until 1996 Nepal had no formal laws in place that either allowed or prohibited foreign investors from requesting foreign jurisdiction. But at the Khimti project negotiators' insistence, in 1996 Nepal amended its Foreign Investment and Technology Transfer Act (FITTA) to legalize foreign and/or multiple jurisdictions for projects involving foreign investment. As if to make things as complicated as possible, the Khimti project used American law for loan documentation, Norwegian law for construction and supply contracts, and Nepali law for the project and power purchase agreements (PPAs) with the Nepal government (Shrestha 2015a).

If the banks had no confidence in Nepal's laws, judiciary, or lawyers, they had even less trust in Nepali companies such as the Butwal Power Company (BPC), Himal Hydro, and Nepal Hydro and Electric (NHE). At bank insistence, the Nepali companies officially became junior partners to their Norwegian counterparts: BPC to Statkraft Engineering, Himal Hydro to Statkraft Anlegg, and NHE to Kvaerner and ABB Energi. BPC, as the project's founder, had initially claimed a 50 percent share in

HPL (the BPC–Statkraft partnership that was developing Khimti). But as negotiations progressed and bank demands became more rigid, BPC found its share reduced to 30 percent, then 25 percent, and finally only 15 percent. (Eventually Statkraft ended up with 73 percent while Kvaerner and the Asian Development Bank [ADB] held the remaining 12 percent.) As one Nepali engineer put it:

> The bankers asked, "Who's building this plant? We have no faith in Nepali contractors. Bring in the experts who have already built 60 MWs." So that's part of the story. I think it was a reality check for Hoftun. Once you get into these huge projects, good intentions alone will not get you through.

Good intentions had brought BPC and its spin-off companies a good distance but they had now met their limit.

Of all the challenges that went into reaching financial closure for Khimti, probably the most contentious—and certainly the most infamous in Nepal today—is the project's PPA: the contract reached between HPL (or the banks that were calling the shots) and the Nepal Electricity Authority (NEA) regarding at what price, and under what conditions, Nepal would buy electrical power from the Khimti plant upon its completion. For any private power producer hoping to sell into the national grid, the PPA is an absolutely crucial factor that makes all the difference between profitability and loss. Private producers want the highest rate possible to maximize returns on their investments. But NEA wants a low rate to minimize costs to itself and to consumers. Of course the best way for a private power producer to maximize profits is to keep construction costs as low as possible. For example, for a new 30 MW power plant, the same purchase rate of, say, 5 cents per unit will be more profitable if the plant is built for USD 30 million than for USD 40 million. BPC began the Khimti project fully aware that the lowest-cost power plant would be both the most profitable for BPC and the least costly for Nepali consumers.

Unfortunately that logic was thrown out the window as international bank financing drove up costs for the final project—from USD 90 to 140 million. With such a high price tag attached, the Khimti project had to negotiate a high power purchase rate simply to pay its annual loan debts. Worse still were the actual rates and terms that the banks commanded for their loans. The World Bank (through its International Finance Corporation [IFC]) demanded an "extraordinarily high" interest rate of 11.48 percent for its loan of USD 31 million while ADB's 31 million also came at the very high rate of 10.5 percent.[7] And both banks demanded that their loans be repaid in a mere eleven years (Pun 2010) rather than the thirty or forty years loans that, according to Hoftun, are common for hydroelectric projects in Norway. In order to repay loans at such high interest rates in such a short period of time, HPL had to demand a very high rate of payment from NEA in its PPA. Of course Statkraft favored a high PPA rate because, after eleven years of debt payment, those very high per-unit electricity rates would become high rates of pure profit.

But if HPL needed a high rate PPA, NEA officials were under less pressure to comply. Nepal needed Khimti's power, but NEA was not ready to sign a blank check for Statkraft and the banks. In March 1994 HPL and NEA (along with representatives from IFC, Norad, and ADB) began negotiations, eventually working through the night to announce on March 28 a PPA of 5.2 US cents per unit of electricity to be paid by NEA to HPL. But word quickly came back from New York (World Bank) and Manila (ADB) nixing the deal. "The IFC just threw it out the window!" said one of the NEA negotiators. "They said it's not a 'bankable project.' Out!" This official remembers a crestfallen Peter Harwood, HPL's general manager, calling on the phone to announce, "I'm closing the [HPL] office. The project is finished, I'm closing down." The banks demanded higher rates (and other concessions) than NEA was willing to pay. End of story, seemingly.

Yet global currents were swirling about in such a way as to allow the Khimti project to reemerge, zombie-like, into the light of day. For the big banks, having the very first private-sector hydropower initiative in Nepal fail sent the wrong message to potential international investors (even if the failure was the banks' fault). The "development" banks were supposedly trying to promote foreign direct investment in places like Nepal—in keeping

---

[7] Another USD 29 million loan from Eskpofinance for electromechanical equipment was at a far lower rate of 5.95 percent but for an even shorter—two year—maturity period (Pun 2010).

with the ideologically driven, business-friendly Washington Consensus that dictated international finance in the neoliberal era (Gyawali, Thompson, and Verweij 2017: 11). With that in mind, one of the Japanese vice presidents of ADB wrote a letter to the Nepal government, encouraging it to reenter negotiations.

In Nepal also the political climate for hydropower development was changing fast. It is here that the Khimti story merges with that of the notorious Arun III project. In the early 1990s the Nepal government had convinced the World Bank to fund a major, 201 MW hydroelectric project on the Arun River in east Nepal. Like a magic potion, Arun III would cure Nepal of its hydropower woes. By mid-decade the project had widespread political and popular support but a small and vocal group of Nepali activists steadfastly opposed the deal. This huge project—with its huge price tag—would, they argued, act like a black hole, sucking into itself all of the national and international hydropower development resources that Nepal could hope to muster for decades to come. Rather than *one huge* project that would almost certainly take decades to finish, involve cost overruns, and be built by foreigners, why not build *many smaller* hydropower projects? Not only would this approach reduce the risks of having all of one's proverbial eggs in one basket, but it would also allow hydropower development to proceed at a scale that Nepalis themselves could build. The anti-Arun III reasoning went thus: rather than one huge foreign-built project, let us have many small Nepali-built projects.

If this logic sounds suspiciously like Hoftun's, that is because it is. Some of the anti-Arun III movement's most important activists were BPC veterans.[8] One of them, Bikash Pandey, had worked as an engineer for

---

[8] In March 1993 anti-Arun III activists formed the Alliance for Energy. One of the Alliance's leaders Dipak Gyawali (writing at that time) argued that Arun III was designed to "serve the need of donors" whereas multiple smaller plants were in Nepal's actual interests (2003: 69). Referring indirectly to BPC's track record, Gyawali notes that, "Despite small hydro generation being a success story in terms of both local manufactures, capacity building, and construction skills, both the World Bank and the International Monetary Fund (IMF) promoted Arun-3 which required major involvement of international contractors and little use of local capacity" (2003: 78). According to Gyawali, the World Bank initially ignored Nepali complaints but, when

UMN's Development and Consulting Services (DCS) in Butwal. "From my perspective," he told me, "as somebody working in the [anti-]Arun campaign, the BPC experience was pretty central to how we envisioned the alternative or why we envisioned the alternative." While Pandey acknowledges other sources of inspiration, he told me:

> For me, [BPC] was an extremely important influence. I mean, I wouldn't have invested time and effort in making this huge argument against Arun III if I hadn't been immersed in that experience and saw clearly that this was possible—that there were things that could be done within Nepal. But the thing is that at that time, we were projecting ahead from very few data points, pretty much all of them from BPC.... You know, build first a 1-megawatt project, then a 5-megawatt project, then a 12-megawatt project, and so on. The point was that this is the kind of trajectory that many companies can take and it's the accumulation of all of those megawatts that has to meet our needs, rather than a single planned project coming every ten years. It's like the tall trees without the undergrowth—nothing to grow after this one is done.

In an article written at the peak of the anti-Arun III agitation, Pandey contrasted Arun III, which he called "a purely extractive form of hydropower development," with projects like Andhi Khola which "are able to make positive, if limited, impacts on the local economy through rural electrification" (1996: 314). Presciently, Pandey wrote, "Arun III and Andhi Khola represent two contrasting categories of hydropower project that are likely to be of great importance in the coming years in Nepal" (1996: 330). Projects like Arun III would leave Nepal dependent, whereas many small Andhi Khola–like projects, built by Nepalis and spread across the landscape, would foster sustainable growth.[9] In the face of this opposition, in 1995 the World Bank withdrew its support for Arun III—the first time

---

the same critiques started coming from prominent western academic and NGO voices, "the Bank could no longer ignore them" (2003: 85–86).

[9] The anti-Arun III forces proved to be correct. Rather than the 201 MW that Nepal would have received from Arun III, in the decades since its cancellation Nepal's public

it had ever backed away from a hydroelectric project because of arguments based in economic development logic, rather than due to environmental or humanitarian concerns.[10]

It is in the context of this post-Arun III political and hydropower landscape that Khimti project negotiations resumed. The World Bank's decision to pull out of Arun III came during a brief UML (Unified Marxist–Leninist) Party government putting UML leaders under huge pressure to launch alternative hydropower development schemes. This meant that, when HPL (and Statkraft, and the banks) returned to the PPA bargaining table, they did so with considerably more bargaining leverage.

The second PPA, signed on January 15, 1996, was a bonanza for HPL and its majority shareholder Statkraft. The deal granted HPL 5.94 cents per unit—*in USD* with annual increases tied to the US Consumer Price Index. In a clause unique to the Khimti PPA, the deal also waived the usual royalty paid on power sold (2 percent for the first fifteen years and 10 percent thereafter), stipulating that NEA must pay the royalty, not HPL. Among other things, this means that, over time, the rate that NEA pays for Khimti power is considerably more than the stated 5.94 cents. To make matters worse (for NEA), because the PPA agreed to pay HPL in USD at a time when USD 1 was worth about NPR 55, today—when the exchange rate is well over NPR 100 per USD—NEA is paying over 100 percent more (in NPR) than it did when the PPA was signed (Butler 2014). As a final provision before they would close the deal—because of the great (perceived) risks involved with doing business in Nepal—the banks insisted on another clause (also unique to the Khimti PPA) called a "sovereign guarantee" whereby if, at any point and for any reason, NEA did not pay for Khimti power, the government of Nepal would be held liable for payment. Thanks to these provisions, today (after its loans have been paid off) HPL makes about USD 35 million per year in profit—paid for, ultimately, by Nepali consumers.

---

and private sector added 499.8 MW of installed capacity at a lower cost than what had been estimated for Arun III (Shrestha 2015b; 2017: 138).

[10] For more on the role of the "Butwal Power Company model" in forming the basis for a challenge to Arun III, see Karki (2017: 120).

Nepalis condemned the second and final Khimti PPA pretty much from the moment of its announcement. Rather than credit for initiating new projects, UML took the blame for caving in to foreign pressure, cheating Nepalis to serve foreign corporations.[11] Ever since, there have been calls to alter the Khimti PPA—either by getting the IFC and ADB to lower their interest rates or to get HPL to return some of its "obscene dividends" to Nepal (Pun 2010).

CONSTRUCTION

After the January 15, 1996, signing of the second Khimti power purchase agreement (PPA), construction finally resumed but now at a furious pace in order to meet bank-imposed deadlines. The final project agreement had included clauses imposing harsh financial penalties on the project's contractors if they did not meet the stipulated four-year completion deadline. Because for Statkraft Khimti was a chance to prove their international construction capacities, they took a no-holds-barred approach to the project. Had Statkraft and the banks not insisted that work stop during financial negotiations, the project might have been much farther along by January 1996. Fortunately, at least some work *had* continued. Part of the Nepal government's contribution

---

[11] Santa Bahadur Pun (2010) points to the interesting circumstance whereby the then minister of water resources was Laxman Prasad Ghimire, a Congress Party member of parliament (MP) from Ramechap district—where Khimti is located. As minister, Ghimire held the unenviable position of leading two units within the ministry of water resources (MoWR) with diametrically opposed interests when it came to the Khimti PPA. On the one hand, the Nepal Electricity Authority (NEA) wanted a PPA favorable to itself and Nepali consumers. But on the other hand, the Electricity Development Centre (EDC)—the MoWR unit charged with promoting private-sector hydropower development—desperately needed a successful implementation at Khimti to justify its existence. "Thus," according to Pun, "when NEA hee-hawed on the PPA negotiation with HPL, the EDC chief is reported to have issued the fatwa that if the Norwegians returned without the Khimti PPA, then 'heads will roll' in NEA. Such fatwa, of course, could not have been issued without the explicit blessing of the supreme Khomeni," namely, Ghimire.

to the project had been to fund a 27-kilometer road from the nearest existing highway to project headquarters, powerhouse access tunnel, and staging area near the village of Kirne on the banks of the Tamakoshi River. Designing and building this road, which includes two major bridges, kept at least some people within Himal Hydro and Nepal Hydro and Electric (NHE) occupied before construction resumed.

Everyone I spoke with emphasized the huge shift in corporate culture that came with the return to work at Khimti. When the Butwal Power Company (BPC) had initiated construction the emphasis was on cost-containment and low-tech solutions. But with Statkraft now at the helm, the logic shifted to big money, big machines, and big hurry. One Nepali Himal Hydro engineer remembered, "We learned a lot but the transition was so sudden. Even after Jhimruk, we weren't prepared for a project like Khimti. It was *so* commercial and the demand was to complete the project at any cost."

This build-it-at-any-cost logic was nowhere more apparent than in the construction of the Khimti headworks. One of the casualties of the long halt in project preparation prior to 1996 was the failure to build an access road from the Khimti headquarters and staging area near Kirne up the Khimti Khola to the headworks location. Money was budgeted but by the time construction got underway, there was simply no time to wait for a road to be built. Instead Himal Power Limited (HPL) invested in a complex transport system involving (human) porters and mules, ropeways, and even heavy-lift helicopters.

During the peak construction period the Khimti project employed over 1,000 porters who, as a group, carried around 2,500 loads per day up and along a steep ridge between Kirne and the headworks site. Because Khimti is not far from the Everest/Khumbu mountaineering or trekking region, the project was able to draw porters to the project by setting its pay rates higher than those in the tourist trade. Furthermore, Khimti offered favorable conditions for porters whereby if they signed on for a one-year contract, they would get a lump sum bonus at the end of the year—essentially a built-in savings plan. This was a major attraction for porters, at least a few of whom were able to use these bonus payments to purchase land. The project set aside land for temporary porter camps but provided no built accommodations. Eventually conditions got bad enough that there were porter strikes demanding better living conditions. Along with large teams of mules, humans carried tons and tons of sand, cement, explosives,[12] nuts and bolts, and other supplies from Kirne along local paths to the headworks. As with earlier BPC projects, Khimti also installed ropeways to transport supplies to major tunnel construction access points (cf. Dhakal 2004).

But certainly the most dramatic answer to the problem of bringing supplies to the headworks site without an access road was the use of helicopters. With the fall of the Soviet Union came the collapse of the Russian–Afghan war and, along with it, the defection of many Russian troops, including pilots who simply took their Soviet helicopters and flew to South Asia. Largely roadless, Nepal was a good market for their services and at Khimti they found eager employers. Able to haul up to three tons per load in these huge, double-rotor choppers, at critical points in the construction process HPL hired small fleets of Russian aircraft that would shuttle twenty–thirty 3-ton box loads of supplies to the headworks construction site per day. This included everything from cement and steel, to massive transformers, to a 40-ton Volvo tank-track excavator that was disassembled into 3-ton pieces, flown to the site, and put back together. By carefully pre-assembling 3-ton loads and then bringing in choppers every few weeks to shuttle loads to the headworks on an hourly basis, HPL was able to get supplies delivered at a relatively low cost. "These guys became *so efficient* at this," remembers one Himal Hydro staffer. But even with experienced pilots, flying heavy loads over a windy ridge and up a gusty, narrow river valley was risky business. One engineer on site remembers heart-stopping moments when rotors lopped off tree branches while trying to settle loads under windy conditions.

The Khimti headworks represented a serious technical design challenge as well as a construction challenge. With Norwegian assistance, engineers at Hydro Lab on the Patan engineering campus took two years and seven full topographic modeling studies before arriving at a design solution acceptable

---

[12] One complicating factor was that, because of the then ongoing Maoist Uprising, or People's War, state security personnel had to be on hand whenever explosives were being used or transported. "Every time you wanted to do blasting, or when you were carrying explosives by mule caravan, you had to have a police guard," remembers one BPC engineer.

to the international lenders. Although it cost almost half a million US dollars, the design has proven to be very effective. One Nepali engineer who worked on the project pointed to the Khimti headworks as a unique and landmark design achievement.

> Environmentally it is so friendly because you don't have a big dam or anything. When you look at it, it's just like a river flowing. It has been working perfectly fine up till now. But the point is that it was a totally new design. It was nothing like taken from a textbook. It was entirely the result of careful modeling and experimenting.

As chief engineer with Himal Hydro, Pratik Man Singh Pradhan spent three seasons (between monsoons) on the intake structure working alongside Dil Bahadur Shrestha, who Pradhan described as a "very brilliant" construction manager trained at Butwal Technical Institute (BTI).

Like Jhimruk, the Khimti headworks were built to international construction standards including the use of super-hard, durable concrete. Porters and mules brought in good-quality Indian cement but what was missing was the high-grade aggregate needed to produce extremely hard concrete. Technicians tested dozens of kinds of rock and gravel from nearby sources but all of them were too soft. Searching farther and farther afield, they finally identified a source of hard rock aggregate but, unfortunately, it was near the top of a mountain some 10 kilometers away! Nevertheless, porters carried tons of aggregate to the site. The scar from the quarry is still visible near a mountaintop visible in the distance high above the headworks site.

Whereas there was deep-seated disagreement between BPC and top Statkraft management over Khimti's financing, relations between the Statkraft subsidiaries and BPC, Himal Hydro, and NHE were much more amicable. Statkraft Anlegg (construction) and Statkraft Engineering sent a range of white- and blue-collar specialists to Nepal where, by all accounts, they formed friendly, cooperative, and productive relationships with their Nepali and United Mission to Nepal (UMN) expat counterparts. In conversations with me, Hoftun stressed that Statkraft went out of its way to select expat workers who would fit into BPC's service and mission-oriented ethos: "no drunks or culturally insensitive types," Hoftun said. Some of the Norwegians were overtly Christian. "They were like part of the UMN group." At its peak, some thirty-five Norwegian engineers, tunnel foremen, and mechanics were working side by side with BPC staff.

Within each company, groups of BPC, Himal Hydro, and NHE engineers worked alongside one or two senior Norwegian engineers from the respective divisions of Statkraft. Almost all of the Nepalis had previous experience at Andhi Khola and Jhimruk but they all reported that working with senior Norwegian specialists was an extremely valuable learning experience. At the top level was the Khimti "executive steering committee" comprised of two Nepalis (Shiva Kumar Sharma representing Himal Hydro and Pratik Man Singh Pradhan representing BPC) and two high-level Statkraft engineers. Every six–eight weeks the Norwegians would fly to Nepal and the four would visit the project site, examining everything—human resources, equipment, materials—to make sure all was in good shape and in stock to guarantee no construction delays. For the Nepalis involved this was an invaluable combination of professional experience and mentoring that prepared them for executive-level management at a scale and complexity they had never before dealt with.

At the white-collar level Nepalis and Norwegians conversed easily in English. But when it came to the more working-class Norwegian foremen and technicians—people who supervised the complex and dangerous work of digging tunnels, handling explosives, operating heavy equipment, and so on—relations with their Nepali counterparts were more difficult because the two sides had no common language to draw on. The Norwegians were highly skilled individuals with deep experience but they had to train and supervise Nepalis through demonstration and hand gestures. When communications broke down more senior Nepali and Norwegian staff would step in to translate but there were repeated tensions between the brawny, no-nonsense Norwegians and the skilled Nepalis they were overseeing.

The Khimti project had elaborate safety protocols in place that were largely up to the Norwegian foremen to follow through on and enforce. For Nepali equipment operators and other skilled laborers (with prior experience at Andhi Khola and Jhimruk) these foremen could appear rigid, rule obsessed, and bossy. Ironically, a series of tragic deaths actually improved cross-cultural relations. Peter Lockwood, a UMN engineer assigned to oversee safety procedures across the entire project site, told me that, all told, six Nepalis died accidental deaths during the construction phase at Khimti, including—about one and a half years into the project—four deaths in a single month: one by electrocution (due to a faulty Indian circuit breaker), one in a rockfall (while not following safety procedures), and two in a fluke explosion. After this terrible month, managers shut down the entire operation while a Norwegian expert came in to review safety procedures. Lockwood described how, up till that time, tensions between Statkraft-appointed Norwegian foremen and Nepali tunnelers had been growing. Yet, poignantly, the Nepali workers changed their opinions at the sight of their gruff Norwegian foremen openly weeping in grief over the deaths of their Nepali charges, people they cared about and were working hard to keep safe in a dangerous environment.

MITIGATION

If the Khimti project enjoyed relatively good relations between Nepali and foreign personnel, fortunately the same was true of relations between project staff and the local community. Having watched the disaster in community relations that unfolded at Jhimruk, at Khimti BPC worked hard to build significant "mitigation" features into the project from the start. It helped that the project had relatively little overtly *negative* impact on local communities: it did not appropriate much local land or affect irrigation or fish. But project developers wanted to make sure that there were *positive* impacts as well. For example, from the beginning BPC planned to hand over a 150-kilowatt (kW) turbine-generator (installed in a tributary stream above the headworks to provide power during construction) to the community for rural electrification.[13] Today a view down the Khimti Valley at night shows thousands of twinkling electric lights from individual homes scattered across the mountainsides, a constellation of glowing dots that seems to blend into the starry sky above.

Like at Jhimruk, at Khimti there was a walled compound with various grades of living quarters for staff, including rather luxurious (by Nepal standards) accommodations for Norwegian Statkraft employees who enjoyed

---

[13] This plant's output was later expanded to 630 kW.

air conditioning, hot water, swimming pool, tennis courts, and manicured grounds. But *unlike* at Jhimruk, at least some of the expat project staff lived in rented rooms and homes right in the nearby village of Kirne. These were mainly UMN appointees (deputed to BPC) who were able to speak Nepali and establish good relations with members of the local community. Many of them had come first during the brief period of preliminary construction and were able to help deal with community concerns during the interim period. For example, when BPC was forced to halt construction it only seemed to confirm rumors circulating locally that the Nepalis and foreigners digging holes in the mountainside were really looking for precious stones or minerals. They were not really building a power plant and they just quit digging when they did not find what they wanted. But positive relations with community members allowed BPC staff to address local concerns and resume construction with local support.

Dealing with community relations and mitigation was new for Statkraft but, at BPC's insistence, HPL applied to Norad (the Norwegian aid agency) for special funds to support community outreach projects. With these funds HPL established a special unit—Khimti Environment and Community, or KEC—focused entirely on mitigation efforts. KEC sent staff to local villages to provide training in things like kitchen gardening and pig farming. KEC supported local schools and, perhaps most significantly, built and equipped a subsidized health clinic at Kirne.[14] Because local people had, from the beginning, associated the Khimti project with BPC, even after Statkraft dominated the project, BPC got credit for the project's positive local impact. One Nepali staff member who had previously been at Jhimruk remembered with amazement how BPC was "very well liked, a very well-trusted company" in local eyes. "Even though foreigners were involved, BPC was seen as a Nepali company." He described how local performers actually composed songs praising BPC for giving them schools, health care, and skills training. They even invited BPC to develop more projects in the area!

> That was the kind of emotion among local people. So it was, in a way, very supportive for us to work in the community. They accept you, they *believe*

---

[14] Clinic management was later contracted out to the Dhulikhel Hospital.

you. Normally what happens is that there's always a conflict between local people and project people which makes it really difficult to come together and do things together.

Like others I spoke with, he noted the stark contrast between community relations at Jhimruk and Khimti. Both projects were located in or near Maoist strongholds during the People's War but while Jhimruk became a favorite target of community hostility, Khimti was untouched. "If we'd had the same disturbances at Khimti, the project never would have gotten done on time." Mitigation efforts were probably a key factor in the project's overall successful, and profitable, completion.

EXCAVATION

In addition to the intake structure, the Khimti civil works included an 8.7-kilometer horizontal headrace tunnel, a 45-degree inclined pressure shaft with 630 meters of head,[15] a powerhouse cavern, a roughly 1-kilometer powerhouse access tunnel, and a 1-kilometer tailrace tunnel. Engineers designed the pressure shaft with a short access tunnel at the midway point so that it could be excavated from four faces simultaneously, two each in the upper and lower sections. Like at Jhimruk, tunnelers used equipment that allowed them to excavate mainly from the bottom up in the incline shafts, thereby allowing gravity to do much of the work of clearing debris. But unlike at Jhimruk, for the first time in Nepal tunnelers used sophisticated "boomer" drilling rigs in the horizontal tunnels. Rather than the handheld drills that could only penetrate about 1.5 meters, these mechanized rigs had two or three drilling booms equipped with 3.8-meter drills that allowed for much greater excavation per dynamite blast. Although much more efficient, the challenge was to keep the complicated boomer rigs working. Some of the Norwegian technicians spent much of their time simply keeping equipment operational.

---

[15] At 630 meters, Khimti holds the record for high-head power projects in Nepal, though, when finished, the Upper Tamakoshi project will surpass it.

At least on paper the Khimti tunneling work was pretty straightforward. The scale was bigger than anything BPC or Himal Hydro had tackled before but, with experienced Norwegian oversite and advanced equipment, the project should have proceeded apace. Unfortunately, preliminary test drilling had failed to accurately predict the actual rock conditions that tunnelers encountered once they were inside the mountain. The lower tunnels and incline shaft were in good, solid rock that presented no unforeseen problems. But in the headrace and upper incline shaft tunnels workers encountered fractured, unstable stone that had to be carefully shored up to prevent cave-ins, thereby seriously delaying progress. Tunnel construction began in the fall of 1996 but by March 1997 workers had managed to complete only 300 meters—most of it lined with concrete and metal reinforcements—of an almost 9-kilometer tunnel. Significantly behind schedule, and facing stiff penalties for failing to meet the December 1999 construction deadline, Statkraft Anlegg and Himal Hydro decided to switch to heavier, more effective construction equipment.[16] Not only did this have to be airlifted in from abroad, but the new equipment required a bigger tunnel diameter—both with serious financial implications. More challenging still was the upper section of the incline shaft. A Nepali engineer working at the site recalled how

> in the upper shaft we were coming close to the breakthrough point when there was a big cave-in, all the way from the surface, like a chimney. There was insurance but still this caused a lot of delay and that really cost us. Himal Hydro wasn't alone but along with Statkraft Anlegg [we had to absorb those losses]. I know this happens in most projects at one point or another. But this was a bigger one, the total collapse of a chimney so that they had to excavate again. There was a lot of delay due to that.

---

[16] Using different equipment obviously did not change rock conditions, but the new machines allowed operators to drill up to 50-meter-deep pilot holes to get a sense of the geology ahead. Whereas with the previous equipment, by the time you knew what you were dealing with, it was too late, the new machines allowed tunnelers to take proactive measures such as pre-grouting or umbrella grouting in unstable sections, thereby minimizing the potential for cave-ins.

According to Hoftun, dealing with Himalayan rock was a big learning experience for Statkraft Anlegg as well. Accustomed to stable rock in Norway, they had not adequately budgeted for the risks of soft rock conditions. By retooling with expensive, new larger-scale equipment and putting it to intensive use on a twenty-four-hour schedule, Statkraft Anlegg and Himal Hydro managed to complete the civil works within the contracted time limit but only at a cost that left them with a major loss against the price they had contractually agreed upon. While Statkraft could absorb these losses, its partner Himal Hydro was in a much more vulnerable position.

For Himal Hydro, Khimti was an important learning experience, but a painful and almost lethal one. Whereas BPC had managed to retain only a 15 percent interest in HPL (the Khimti project's official owner or developer), Himal Hydro had negotiated to shoulder 35 percent of the project's civil works construction contract along with its majority partner, Statkraft Anlegg. This meant that Himal Hydro would enjoy 35 percent of the profit earned on the construction contract or—as it turned out—be responsible for covering 35 percent of the project's cost overruns. One engineer I spoke with estimated that the Khimti construction consortium (Statkraft Anlegg and Himal Hydro) lost around USD 1,000 per meter in the project's headrace tunneling work (around USD 8 million) and between USD 10 and 12 million overall.

As general manager of Himal Hydro, Shiva Kumar Sharma watched in horror as project expenses quickly went over budget. "I was *very worried* because if we lose 10 million dollars, then 35 percent of that—3.5 million dollars—well, that was more than the *whole capital* of Himal Hydro! It was very tough." Already in April 1998 Sharma received a letter from Statkraft Anlegg management warning Himal Hydro—in "rather harsh language"—of its impending bankruptcy if Himal Hydro did not immediately refinance the company. Fearing that the letter was a threat from Statkraft to drive Himal Hydro into bankruptcy and then take over, Sharma wrote to Odd Hoftun asking him to look into the matter. After speaking at length with Statkraft Anlegg staff in Norway, Hoftun reported back to Sharma[17] that Statkraft

---

[17] In an unpublished letter dated April 20, 1998.

had no intention of seeing Himal Hydro fail, an outcome that would in fact be bad for Statkraft's own commercial interests. Rather, they were warning Himal Hydro not to go into its scheduled annual shareholder's meeting reporting a loss that would invoke panic and potentially have negative consequences for Statkraft as well. Better to postpone the meeting until Himal Hydro's finances could be shored up. To that end, Statkraft offered to sink new investment capital into Himal Hydro to the point where it would control 49 percent of the company's shares. They would become a major, but not majority, shareholder in Himal Hydro. In the end UMN was more than willing to divest itself of Himal Hydro stock, leaving Statkraft as, in fact, the majority owner.

Statkraft Anlegg's intervention in Himal Hydro's affairs was partly the result of the goodwill that existed between Norwegian and Nepali staff. Having worked side by side with Nepali colleagues—and, in a way, being responsible for underestimating the project's risks—Statkraft had no interest in seeing its pioneering Nepali counterpart fail. But there were also more hardheaded financial reasons for Statkraft to save Himal Hydro's skin. As the following chapter will make clear, by the late 1990s a new large-scale project (Melamchi) was under negotiation with preliminary funding contingent upon the project being undertaken by Nepali companies. Because Himal Hydro was the only Nepali civil contractor in contention for this new project contract, it was in Statkraft's interest that Himal Hydro survive, with Statkraft as a major shareholder. Bailing out Himal Hydro would give Statkraft Anlegg a firmer foothold in the potentially enormous Nepali hydel development market.

For NHE, Khimti was a more unequivocally positive experience. Like Himal Hydro's relationship with Statkraft Anlegg in the Khimti civil works contract, NHE was junior partner with Norwegian turbine manufacturer Kvaerner in the Khimti electromechanical equipment contract. Just as they had at Jhimruk, NHE fabricated all of the project's large-scale hydraulic components—penstock, bifurcation tubes, valves, turbine casings—except for the inner Pelton turbine blades or "runners" (equipment that Kvaerner held proprietary control over).

Even long before the Khimti finances were finally nailed down, Hoftun knew that *if* the project went forward, NHE would need a new manufacturing facility in Butwal to actually build components at the scale that Khimti would demand. To that end, and as part owner of NHE, BPC invested its own scarce resources in building a new workshop for NHE in Butwal. In Norway Hoftun sat down with an experienced Kvaerner engineer to draw up plans for a facility that would allow NHE to build the kind of *big* equipment that future hydel projects in Nepal would demand. The new NHE shop featured extra-large bays with two heavy-duty 8-meter rolling cranes together capable of lifting 60 tons. Fortunately this new facility was in place by the time the Khimti project got the go-ahead.

As they had for the equipment built for Jhimruk, NHE used skilled Nepali welders and machinists working under Norwegian supervision to produce components that met stringent international quality standards. At critical points Kvaerner sent personnel to Butwal to give specialized training and oversee quality control. From Butwal the penstock tube sections went by truck and then ropeway to the Khimti pressure shaft where they were lowered into place, welded together, and encased in concrete. I left a visit to the Khimti powerhouse in awe of the huge bifurcation tubes, valves, casings, and other components that were made right there in Nepal. According to Hoftun, one of Khimti's most important legacies "was the long-term impact that the project had in building up professional morale and pride among staff and workers in NHE, inspiring them to maintain and strengthen the good reputation of NHE."

CONCLUSION

By the time Himal Power Limited (HPL) began commercial operations at Khimti on July 11, 2000,[18] the project was already one for the history books. With significant help from the Nepal government, Khimti had pioneered the route to private-sector power development in Nepal. New legislation in 1992 had initiated the project exploration, an unprecedented power purchase agreement (PPA) in 1996 closed the deal and launched

---

[18] The project's official, ceremonial inauguration at the hands of King Birendra came on November 27, 2000.

construction, and further policy implementation in 1998 provided legal assurances that smaller-scale hydropower developers needed to lure them into the market.[19] By demonstrating that private development of Nepal's hydropower potential was both possible and highly profitable, Khimti encouraged Nepali and foreign capital to try to fill the growing chasm between Nepal's power supply and demand. Today's thriving independent hydropower sector—much of it driven by Nepali professionals, skilled workers, and capital—would have been impossible without these developments. Odd Hoftun's dream of an indigenous Nepali hydropower development sector was becoming a reality.

Yet for the Butwal Power Company (BPC) the Khimti project phase (roughly the decade of the 1990s) was a disastrous period. BPC launched Khimti from a position of dynamic growth, strong Nepali participation, and a very positive team- or family-like corporate ethos—only to see its vision of Nepali human capacity building ousted by a global commercial agenda of profit extraction. Hoftun had wanted to prepare BPC and its daughter companies for the harsh realities of market competition but in the process—perhaps inevitably—the humanitarian and nationalist vision that had brought them thus far gave way to the logic of the market.

Further complicating this shift was BPC's transition to state ownership in the fall of 1996, about the time that construction at Khimti began in earnest. As outlined in the previous chapter, what was to have been a controlled handover of Jhimruk to the Nepal Electricity Authority (NEA) management suddenly escalated into the wholesale transfer of BPC itself into government control. For most of the Khimti construction period, BPC was a state-run enterprise with board control no longer in Odd Hoftun's hands. BPC survived, basically running on momentum built up over the previous decades. But by 2000, the state's failure to line up further development work for BPC (admittedly a difficult task at the height of the People's War) led to a steady decline in company morale and manpower.

---

[19] In 1998 the Nepal government passed legislation requiring the Nepal Electricity Authority (NEA) to purchase power from IPPs at a fixed rate for up to 5 MW and a negotiable rate for 5–10 MW plants (Karki 2017: 120).

The next chapter examines Odd Hoftun's desperate efforts to keep BPC, Himal Hydro, and Nepal Hydro and Electric (NHE) alive by linking their fortunes with the proposed Melamchi project. But none of those hopes materialized soon enough. After Khimti, Himal Hydro found itself with large amounts of sophisticated equipment—from excavators and cranes to boomers and ropeways—but no work for them to do. "Well," said one former Himal Hydro engineer, "there's a yard in Patan where all that equipment is stored. Himal Hydro used some of this equipment for some of their little contracts here and there. But most of the equipment, like the entire ropeway, that's still lying there."

Maybe even more valuable were the human resources that companies like Himal Hydro found themselves with at the end of Khimti. From engineers to heavy equipment operators to mechanics to construction workers, Khimti had employed and trained hundreds, maybe thousands of skilled Nepalis. But with little or no demand for their skills at home, these people soon looked abroad. A former Himal Hydro administrator described how most of these men ended up in the Middle East, forming one of the early waves of Nepali labor migration to the Gulf.

> These operators, their skills were in so much demand in the market that most of them just left. Because at Khimti there was so much mechanization—whether it was the ropeways or the boomers or the crane operators or excavator operators—they were trained in such a professional manner. But after Khimti we didn't have the projects that we could build using all that trained manpower. So there was a sudden gap. And people can't just wait. But there was demand in the Middle East and once they had a few connections, then *snap!*—everybody took off. Even today, most of them are still there, making good money.

Another former Khimti engineer told me of Nepali technicians and managers from Khimti who found work in Europe and the US. One of the Khimti tunnelers is now supervising construction on the new high-tech Elizabeth line of the London Underground (subway) system.

Ironically, the first international direct investment in Nepal's hydropower sector, backed by major "development" bank financing, helped spark an *exodus* of Nepali labor power—both professional and skilled technical workers—from the very sector that Hoftun and others had worked to build up. This pattern continued in the coming years as the same international banks helped to wrestle control of the Melamchi water diversion scheme away from BPC and into the hands of foreign developers. More than twenty years, and almost half a billion US dollars, later, the Melamchi project is still unfinished.

6

MELAMCHI AND THE RUSH TO PRIVATIZATION

Odd Hoftun may have given up his official responsibilities in Nepal when, in 1994, he resigned his post as general manager of Himal Power Limited and retired to a tidy cottage in his wife's home village along the south coast of Norway. But "retirement" proved only to be the next phase of Hoftun's active career. Hoftun simply could not sit and watch the Butwal Power Company (BPC)—that he and so many others had nurtured for decades—sink into oblivion, especially after its nationalization in 1996. Once in government hands, it was as though BPC was placed on a shelf to be exploited occasionally but mainly ignored and allowed to disintegrate. For BPC simply to survive, let alone continue its growth curve along with its allied companies, it needed two things: new work and new management. It needed a new large-scale hydropower construction project to develop, and it needed to be brought under private ownership that would work to further its original mission of human capacity building in the interests of Nepal.

In what was to be the last major campaign of his career, Hoftun—along with others who shared his concerns—set out to accomplish these two goals. They would secure funding for the Melamchi Diversion Scheme (MDS) that would guarantee work for BPC, and they would spearhead a coalition of investors that would purchase BPC and allow them to again exert significant influence at the board level. Broadly speaking, both of these challenges were eventually met: today the Melamchi tunnel is almost completed, twenty-plus years after Hoftun's push, and BPC is a private company in Nepali hands. But neither of these efforts turned out as Hoftun hoped. BPC was eventually elbowed out of the Melamchi project by international contractors supported by so-called development banks. And the privatization process that Hoftun hoped would lead to Norwegian majority control of BPC eventually devolved into a nightmare of bureaucratic obstructionism that finally ended—six years after it began!—with a Nepali ownership whose priorities were often not Hoftun's.

Once in "retirement"—but now no longer tied in any formal way to the United Mission to Nepal (UMN) and its priorities—Hoftun resumed his place at the chessboard of modern Nepali history attempting to outwit his opponents, channel momentum, and maneuver his pieces into positions of strength. This chapter and the next follow this process from roughly the mid-1990s to 2004 when BPC's privatization finally concluded. Because the Melamchi project and the initial push to privatize BPC are so strategically linked, this chapter tells their stories simultaneously.

THE MELAMCHI DIVERSION SCHEME

In the early 1980s Shiva Kumar Sharma, then a junior engineer with Himal Hydro (and later its general manager for eighteen years), began advocating a bold idea. Already then the Kathmandu Valley had a serious drinking water shortage. But Sharma, fresh from a six-year course in mining engineering in Russia, also had tunnels on his mind. Drawing on earlier survey and engineering studies (Gyawali 2015: 12), Sharma published an article in *The Rising Nepal* promoting a scheme whereby the Melamchi River, which flows at a higher elevation in valleys north of Kathmandu, would be diverted through a roughly 30-kilometer tunnel under the Shivapuri Mountain and into Kathmandu's water supply. With the Nepal government and foreign developers beginning to show interest, by the late 1980s the World Bank and the United Nations Development Program (UNDP) funded a series of pre-feasibility, environmental, and full-scale feasibility studies but the scheme seemed too costly and difficult for anyone to actually undertake.[1]

In 1994 engineers with the Butwal Power Company (BPC) Hydroconsult, who were then drawing up plans for the Khimti project, proposed a different

---

[1] Much of this Melamchi narrative draws from two unpublished documents by Hoftun (2007, 2013).

approach to Melamchi. Rather than simply a long, costly water tunnel, why not combine the water delivery scheme with a hydroelectric generating plant? Not only would this accomplish two goals in one project (water *and* power—both desperately needed), but one would essentially pay for the other. According to this plan, the commercial power-generating component (similar to Khimti) would attract private investment which, in turn, would pay for all or most of the tunneling cost. Once through the turbines, the water would be delivered (free or at least steeply subsidized) to the north side of the Kathmandu Valley. Furthermore, the plan would save money by following the "Jhimruk model"—namely, using Nepali companies (BPC, Himal Hydro, and Nepal Hydro and Electric [NHE]) and relatively low-cost construction techniques. Initial estimates put the price tag for the tunnel *and* a 25-megawatt (MW) hydropower plant at around USD 100 million. Based on a BPC Hydroconsult concept proposal, UNDP agreed to fund a new, more detailed feasibility study to be carried out by BPC with foreign technical assistance.

In February 1996 BPC took this newly completed Melamchi "bankable feasibility study" to the Nepal government. At the height of the winter dry period—when both water and power are at their scarcest—it did not take much to convince officials of the plan's merits. The government established Melamchi Water Limited, a public–private company charged with undertaking the Melamchi project, and immediately instructed BPC to take the lead as the project's principal private equity investor or developer. (Ironically, all of this happened in the midst of the United Mission to Nepal's (UMN's) transferal of BPC ownership into government hands, making things all the more complicated.)

For BPC's general manager P. P. Adhikari, already busy managing the contentious and complicated work then underway at Khimti, having the Melamchi project dropped in his lap was simply too much. In desperation, in April 1996 he called on Odd Hoftun—his former boss and informal advisor—to ask if he would assist in managing BPC's involvement with Melamchi. Hoftun was torn. On the one hand, he was eager to keep a hand in BPC affairs and eager to support his friend and colleague Adhikari. But on the other hand, fresh from stinging experiences surrounding the Khimti project's financial closure (where outside commercial interests had hijacked the project), Hoftun was wary of the challenge. Ultimately he agreed to help out in the hope that, based on its recent experience, BPC might be able to avoid the entanglements that had pulled the Khimti project out of their hands. The other powerful motivating factor for Hoftun was the simple knowledge that, if BPC and its collaborating offshoots were to survive, they would need a major new project set to commence in four years when Khimti was finished. The futures of the companies, as well as thousands of skilled workers, depended on getting the Melamchi scheme nailed down.

Hoftun's main job was to shop the Melamchi feasibility study around to potential donors and investors. Not surprisingly, he turned first to the Norwegian development establishment that had so faithfully supported Hoftun's development vision in the past. Coincidentally, based on these earlier relationships, Norway had recently designated Nepal as one of ten countries in which it would focus its foreign aid work. In Oslo Hoftun pitched the BPC Melamchi plan to Norwegian aid officials, suggesting that this project would be an excellent way to put its new policy into practice. They agreed and, after negotiations in Nepal, in early 1997 Nepal's minister of finance formally applied to Norad for money to launch the Melamchi project.

In return for preparing proposals and shepherding them through complicated political structures in Norway and Nepal, at this point Norad offered to hire Hoftun and a Nepali colleague as consultants. Norad covered Hoftun's travel expenses. For a salaried coworker Hoftun turned to his old friend and BPC colleague Balaram Pradhan. Like Hoftun, Pradhan was distressed at the current state of BPC affairs—with its power plants poorly managed and little effective leadership from the government—and saw Melamchi as a way of saving the company. Speaking of himself and Hoftun, Pradhan told me that "we had shed blood and sweat for BPC. I wanted to help." As they had before, the two made a good team. After decades of work, Hoftun had good contacts in government and other circles, both in Nepal and Norway. But Pradhan, outgoing and sociable by nature, had an unbelievable, almost encyclopedic, familiarity with the "who's who" of Nepal—in addition to his intimate knowledge of the hydropower scene. (At one point later in their collaboration, after Pradhan had described a series of meetings and personal visits, Hoftun asked in wonder, "Is there no end to your connections?") Hoftun assumed responsibility for developments on the

Norwegian side, but the bulk of the work fell to Pradhan who had to guide the proposal through complicated (and ever-changing) bureaucracies across multiple government ministries (finance, water resources, public works, and others).

In March 1997 Hoftun prepared a "Three Component Package Proposal" for BPC to present to the government. The "package" included: (1) a river diversion tunnel and hydropower component to be privately developed with grant funding and commercial investment, (2) a bulk water supply scheme funded by the World Bank and the Asian Development Bank (ADB) that would completely upgrade the Kathmandu Valley's aging (leaking) water and sewage pipes, as well as build a new water treatment facility, and (3) a new retail water distribution and management system for the valley. Any one of these three initiatives would have been a major undertaking but together they added up to an estimated USD 272 million to accomplish. For any one of them to be feasible, the other two had to be built. But since parts 2 and 3, and half of part 1 had to do with *water supply alone*, the only extraneous element seemed to be the hydropower scheme. This proved to be the Achilles' heel for BPC's involvement in the grand scheme.

Since the BPC story only figures directly into the first of these three components, and because the second and third each have their own unique political dynamics, in what follows I will focus mainly on the tunnel and hydropower project.

Through the summer of 1997 Pradhan and Hoftun walked the "three component" proposal through the halls (and sitting rooms) of power in Kathmandu, lobbying for its merits over a series of other related proposals, including one from the ADB that would do away with the project's hydropower element. At a September 9, 1997, meeting, headed by Prime Minister G. P. Koirala himself, the assembled ministers and other top officials chose the "three component" proposal on the condition that Norway would provide initial funding. The Norwegian ambassador promptly complied with an offer of NOK 70 million (around USD 11 million) to get the project rolling. Part of this money authorized BPC Hydroconsult (along with the Norwegian consulting firm Norplan) to carry out a "final design study" for Melamchi, including a high tunnel that would drop water down a shaft into a hydel plant.

ADB's plan had not been selected but they were not easily dissuaded. A few weeks after the September 9, 1997, meeting, ADB officials phoned the Norwegian embassy offering to join as project funders with a contribution of USD 80 million, but on the condition that ADB take control of the project. This, in turn, set off another marathon period of intense lobbying as BPC and Norplan promoted the plan *with* a hydropower component, while ADB and their consulting firm, Nippon Koei, pushed hard to *remove* power generation from the grand scheme. It was another two and a half years before the chips finally fell.

PRIVATIZATION

While the great powers of international development debated the Butwal Power Company's (BPC's) future with regard to the Melamchi project, another crucial dimension of the company's fate was under discussion in Kathmandu. The Nepal government's decision to take over BPC in 1996 had been controversial from the start. No one in BPC wanted it to happen. It starkly contradicted rhetoric coming from international development entities who strongly promoted the *privatization* of government assets, *not* their seizure from the private sector. It even went against the Nepal government's own officially stated planning policies. Therefore, almost as soon as the government nationalized BPC, there were calls from every quarter to privatize it—that is, sell it off to the highest qualified bidder.

In a series of developments that seems anything but coincidental, during the summer of 1997—while Hoftun and Pradhan were vigorously selling the Melamchi "three component package" to government officials—Ram Sharan Mahat, the minister of finance, officially set in motion plans for BPC's privatization. With World Bank funding, Mahat established a "Privatization Cell" within the ministry whose staff included (officially as "advisor") Douglas Clarke, a legal expert from the British Adam Smith Institute. The cell's job was to oversee the privatization of government-owned commercial assets—including several Soviet- and Chinese-built factories, the national airlines, and BPC—and to make sure the bidding and sales process went forth with a high degree of transparency and in strict accordance with international law.

Crucial to note here is the fact that BPC's participation in a Melamchi hydropower development scheme almost demanded that the company be privatized. BPC was to spearhead the project as developer but the plan's success hinged on BPC's ability to attract the private commercial investment needed to pay for the hydel component and its long tunnel. Hoftun, Pradhan, and (probably) Mahat knew that no major investors were going to put money into a project run by a government-owned company, especially a government like Nepal's that was known for its (at best) instability and inefficiency or (at worst) corruption and civil war-induced chaos. For the BPC-backed "three component package" to actually work, BPC would need to be in competent private hands in a way that soothed international investors' fears about investing in Nepal. That is, if Nepal wanted to get the Melamchi project up and running fast, it needed to get BPC privatized fast (so that BPC could bring in the private capital needed to finance the project).

In this way Melamchi was the key to Hoftun's goals of keeping BPC alive (by providing it with a major new project to begin after Khimti) and regaining corporate control by leading a privatization bid. The government's sense of urgency for doing something to address the Kathmandu Valley's water and power shortage added powerful momentum to the push to privatize BPC.

Hoftun and Pradhan lost no time in organizing a privatization bid for BPC. An amended version of the earlier (March 1997) "three component package" proposal, dated November 1997, suggests just how much they had already accomplished (Hoftun 1997). First of all, Hoftun reports optimistically about BPC's impending privatization. "The process of privatizing BPC is being pushed ahead vigorously by the Ministry of Finance, and it is expected that invitation for tenders will be issued by February or March 1998." In fact, reports Hoftun, the Nepal government "is committed to complete their part of the process within the current fiscal year, i.e., before 15 July 1998" (Hoftun 1997: appendix IV, 4, 7). In only half a year, BPC's privatization would (supposedly) be a done deal.

A second key development Hoftun mentions in this November 1997 report is Norwegian support for BPC's privatization. By this time Norad had designated NOK 70 million for the "three component package" but here Hoftun reports that *Norad funding has been made conditional on at least two points*: that Nepali corporations be used to the maximum extent possible in the Melamchi construction process, and that those Nepali corporations be privately held, indicating that—because BPC is the only contending Nepali developer—BPC must be privatized before it takes up the project. Hoftun writes that Norad's position "strongly indicated that the private developer of Melamchi should be BPC under new and competent private ownership" (1997: appendix IV, 5). Making Norwegian aid contingent on BPC's privatization provided Hoftun and Pradhan powerful leverage in their efforts to push privatization through and, hopefully, in their direction.

Finally, this November 1997 report mentions, for the first time, that Hoftun and Pradhan had already mobilized at least the nucleus of a group of foreign and Nepali investors that planned to make a bid to purchase BPC.

> A consortium has now been established and are preparing themselves for participating in the bidding.... This group of potential investors includes foreign utilities who are well acquainted with the development and operation of hydropower plants as well as with retail distribution and consumer services. Their combined financial strength should be more than sufficient to back up BPC in the future. (Hoftun 1997: appendix IV, 7–8)

A few years later, Hoftun wrote:

> When it became clear that HMGN actually was serious about selling a majority share of BPC to private investors, my concern was that there would be at least one qualified bidder with good intentions. So I personally started looking for potential investors who would fulfil these criteria: utilities with necessary technical and financial strength prepared for long term involvement, along with genuine Nepalese partners willing to participate in national development through sound and clean business practice. The result was Interkraft Nepal.[2]

---

[2] From an undated memo (probably around 1999) signed by Hoftun entitled "To Whom It May Concern."

Exactly whose understanding of "good intentions" would prevail in this consortium remained to be seen, but clearly Hoftun intended them to be his own.

Interkraft Nepal (IKN) was the result of Hoftun and Pradhan's quick work to mobilize Norwegian and Nepali investor support for a BPC bid. Though still very much a work in progress, the basic framework of the consortium that would bid for BPC was already there, involving both Norwegian and Nepali investors.

The consortium's name, Interkraft Nepal, came from the main Norwegian investor, Interkraft, a collection of five municipally owned public utility companies in southern Norway that had banded together to improve efficiency, service, and profitability. Together they owned and operated 55 power plants in Norway (ranging from 1 to 328 MW) with a combined generating capacity of 2400 MW,[3] as well as managing a large distribution and retail sales network. If one of the main conditions for the BPC privatization bid was that the bidders have extensive experience in hydel development and management, the Interkraft group easily met those qualifications.

But why would a group of big successful Norwegian utility companies be interested in buying small and struggling BPC? The answer is one that we have heard already from other Norwegian entities involved with this story: Norway had already met the limits of growth within its domestic hydropower market. There were essentially no more projects to develop in Norway, leaving a vast economy associated with hydropower development with little to do. For Interkraft, BPC represented a useful toehold in the Nepali market for hydropower development. The fact that BPC already had a history of close ties with Norway was also appealing. In a memo from 1999, the director of IKN lays out why the Norwegian utilities were interested in BPC.

> One of the main reasons was that BPC was raised by heavy economic contributions from Norad and much Norwegian knowledge is grown

---

[3] If one adds statistics from the power plants that Interkraft partially owned, the total output numbers are much higher.

into the company. Further, IKN see the interest of using BPC as a tool to participate in the development of the huge hydropower resource of Nepal and that we, through such an engagement, could maintain our highly-developed knowledge in construction and operation of hydropower plants. An engagement in BPC, which could give development of highly needed infrastructure in Nepal through a healthy economy of the company, fitted ideally to our strategy.[4]

The same document explains that the Norwegians had every intention of making substantial profits on their investment—between 16 and 20 percent—but that they were taking a long view and those profits would come later only after BPC had grown substantially.

Our true intention is that we want to stay by BPC on long term basis. The need for repatriation of our—the involved power companies—investment is at least for the coming years not strongly needed and we want to promote this idea between the Norwegian contributors. Hence, we will try to use BPC as the appropriate tool and the cash generated by BPC operation to invest in more good projects in Nepal, both for Nepal's benefit, for the Nepali investors' benefit, but also for our benefit, re. the above-mentioned background for engaging in Nepal.[5]

At least on paper, and at least for a while, the Norwegians were backing Hoftun's commitment to plow profits back into BPC.

Finally, a significant motivating factor for the Norwegians seems to have been the Melamchi project itself. Aware of initial Norad funding that essentially guaranteed BPC's participation in a major development project, becoming majority owners of a privatized BPC would allow them to hit the

---

[4] From a document entitled "MEMO on the Situation for the BPC Rebidding Process on 6 Oct. 1999 as Seen from Interkraft Nepal AS," 1, signed by Genne J. Hegglid, director of the IKN board.

[5] From a document entitled "MEMO on the Situation for the BPC Rebidding Process on 6 Oct. 1999 as Seen from Interkraft Nepal AS," 4, signed by Genne J. Hegglid, director of the IKN board.

ground running. The "business plan" appended to an undated, unsigned "Summary of Norwegian/Nepali Bid for BPC" states:

> The bidders are ready to invest in the development of the Melamchi Hydro Power Project as an Independent Power Producer (IPP) as foreseen in the preliminary agreement between NORAD and HMGN concerning funding of the Melamchi Water Supply Scheme for Kathmandu.

In other words, beyond just investing in BPC, the Interkraft investors saw their involvement in Nepal as a chance to invest in actual privately developed power plants—which was, after all, where the real money was to be made. Profits could be plowed back into BPC but, as the Khimti power purchase agreement (PPA) already suggested, profits could be extracted from the power plants themselves.

In an email to Balaram Pradhan in early January 1998,[6] Hoftun reported on earlier meetings with the Norwegian group at which "the necessary decisions were taken in a very satisfactory manner." "But now the Nepali group needs to catch up," Hoftun told Pradhan. "Please follow up on this." At the end of January representatives from Interkraft would be arriving in Nepal (with Hoftun) with the aim of finalizing a memorandum of understanding between the two sides.

Pradhan's job was to pull together the Nepal side of the IKN collective. (Even though Pradhan [like Hoftun] was officially employed as a Norad consultant on the Melamchi project, the BPC privatization effort quickly became a major part of the job.) Although Pradhan had deep experience in Nepal's hydro sector broadly and BPC specifically, he was himself not interested in becoming an investor. Instead, he turned to an uncle, Gyanendra Lal Pradhan, who was then already a successful businessman and owner of several small manufacturing concerns in Nepal. According to Tara Lal

---

[6] January 1, 1998, marks the beginning of a long and detailed email correspondence between Hoftun and Pradhan. Both men were just learning how to use the medium but both agreed that it was better than trying to rely on poor phone connections and faxes. Because both kept copies of their correspondence (which they shared with me), the move to email also left a very useful historical record.

Shrestha (Balaram's biographer [Shrestha v.s. 2074], who had interviewed Gyanendra Lal, and whom I later interviewed), one day toward the end of 1997 Balaram showed up at Gyanendra Lal's office asking to see him. Though annoyed at being interrupted by someone without an appointment, Gyanendra Lal agreed. Balaram walked in and "without any formality said, 'Let's buy BPC!'" At this Gyanendra Lal simply laughed. He knew nothing about how power could be produced and sold to the government, only about buying and selling goods on the market. "How can you buy and sell things that aren't buyable and sellable?" he asked Balaram. But after a quick lesson in the economics and feasibility of the matter, Gyanendra Lal found himself getting more and more interested, even excited, at the prospect.

After some preliminary calculations regarding costs and benefits, according to Tara Lal Shrestha, Gyanendra Lal "came to the conclusion that, if Balaram Pradhan—who knows BPC in and out—helps him, he is ready to buy *all the shares*." The main problem with the plan was that although Balaram offered to advise him informally, he would not join any formal, legal business alliance. This caused Gyanendra Lal much anguish because, though very tempted at the prospect, he was also afraid to go it alone without the invested participation of an industry insider like Balaram Pradhan. After losing sleep over the matter and calling it one of the most difficult decisions of his life, Gyanendra Lal Pradhan decided to hedge his bets and spread the risk by inviting other Nepali Newar businessmen to join a collaborative effort to buy BPC.

Balaram Pradhan agreed to work with Gyanendra Lal in recruiting additional investors, explaining the hydropower sector to many interested parties. As things stood, the government was selling off 75 percent of BPC shares. The Norwegians planned to purchase a controlling interest at 51 percent. The Nepali coalition would purchase the remaining 24 percent. The cost per share was still unknown but with the deep-pocketed and experienced Norwegians leading the way, going in on the BPC purchase made sense. Plus, Balaram told them, this was not only good for business, it was also good for the nation: BPC was using Nepali human resources to build Nepal.[7]

---

[7] According to Tara Lal Shrestha, one of the Pradhans' motives from the start was to keep BPC out of Indian hands or, even worse, the hands of Nepal's Marwari business

Early 1998 was a very poor time to be rounding up investors. The US-led global stock market crash and subsequent depression was not making anyone feel bullish. In a January 20 email to Hoftun, Balaram Pradhan lamented that many of the potential investors he spoke with are "very much discouraged for new investments because of the financial slump." Nevertheless, a core group of Nepali business houses (Newar owned) emerged, its numbers fluctuating but eventually totaling seven (including Gyanendra Lal Pradhan). From their perspective joining the bid was a risky proposition. The global business climate was bad, none of them knew anything about hydropower, and the prospect of forming a joint venture with other, often competing, Nepali business houses was not at all reassuring (big egos, big potential conflict). But they were interested nonetheless.

In an interview, one of the leading Nepali investors in IKN told me of some of the rationale that went into the Nepali group's decision. Members of the group were into many kinds of business, none of them hydropower related, but they all agreed that electrical *power*, or more specifically the lack of it, was an overwhelming concern that all of them shared.

> When we got together we said, "Look, hydropower is a sustaining business." Because we were all in construction and technology and manufacturing. And we all felt that, *the need for power*. I mean, like *every day*! Our factories were running on diesel. We were having huge issues with power so we all decided [to join forces in the bid].

Individually none of them was ready to take on the risk. But collectively, with risks and responsibilities spread, and with the Norwegians taking the lead, the Nepali investors decided that this was a good opportunity to get into a business sector that was bound to grow, but do so in a way that would not disrupt their other ongoing businesses. Furthermore, they very intentionally set the ground rules for their partnership: all would invest equally, no one would have more of a voice than others, and all decisions

---

community. As we will see, this definitely became a motivation later when the IKN group found that their main competitor for the BPC bid was the Chaudhary Group. But at the end of 1997, this revelation was still years in the future.

would be made collectively. They chose the name Shangri-La Energy Limited (SEL) for their collaborative venture.[8] By some ways of reckoning, these were the first Nepali investors in Nepal's still embryonic private hydropower sector.[9]

In a foreshadowing of troubles to come, Gyanendra Lal Pradhan quickly took the lead as the group's chief representative and lobbyist. Though he had no formal leadership role within the SEL group, because Pradhan was proactive and willing to invest his time, the rest of the group was happy to watch over his shoulder. It fell to Balaram Pradhan to act as the liaison and mediator between the Nepali investors and Hoftun and the other Norwegians. It was Pradhan also who had to deal with an ever-changing array of often obstructionist government officials and bureaucrats. Hoftun was not always happy about developments but, from his place on the sidelines in Norway, he could maintain his position as an idealist. It was Pradhan who had to work in the realm of reality, honing his skills at the art of the possible.

THE FIRST PUSH TO PRIVATIZATION

At this point the stories of the involvement of the Butwal Power Company (BPC) in the Melamchi project, and of BPC's privatization, become so intertwined that they have to be told together. Each seemed to be contingent upon the other and Hoftun and Pradhan seemed to have strong backing from officials on both the Nepal and Norway sides pushing to see things move ahead quickly. But there were also formidable forces lining up in opposition to both prospects. Elements within the Nepal government, especially the Nepal Electricity Authority (NEA) and the ministry of water resources (MoWR), had no desire to lose control of BPC through privatization,

---

[8] It took some time for the Nepali investors to formalize their relationship. In January 1999 they agreed to form a joint-venture company.

[9] When the BPC privatization process began in 1998, there were no other Nepali private sector investors. Yet, ironically, by the time BPC's privatization finally concluded in 2003, there were already several other private Nepali hydel companies up and running.

regardless of what various ministers and international agencies called for. With regard to Melamchi, powerful interests wanted to remove the entire hydropower component (and therefore BPC) from the project. And, perhaps most ominously, the Maoist Rebellion (or People's War) was gathering steam and throwing the Nepali state into greater chaos than ever before. Selling the hydropower project and keeping the state focused on the privatization process were full-time jobs for Pradhan and Hoftun that only became more difficult as time passed.

On the Melamchi front the main debate was whether or not to include hydropower generation in the total package. The Norwegians favored hydropower and BPC's participation in the project but their contribution of only about USD 11 million meant that the project needed much larger donors as well. The Asian Development Bank (ADB) wanted in, but on its terms. Already in January 1998 Hoftun complained that "ADB is behaving in a very arrogant matter" that had Norad officials, and Hoftun, up in arms.

What ADB objected to was Melamchi's hybrid character. With both a public water delivery and a private power generation aspect, it was (to them) neither fish nor fowl. ADB's authority on such matters, a senior expert who had overseen water delivery projects elsewhere, strongly objected to the mixing of private commercial interests with what should be a publicly available resource—water. Public water and private power were incompatible and would interfere with each other, ADB argued (even though, as Hoftun pointed out, the two had often been successfully combined in Norway). Was this a water system with a power plant or a power plant with a water system, they asked. And which ministry would take charge? The ministry of housing and physical planning would control the water system, but MoWR would take on the hydropower system. Mixing the two would be a bureaucratic disaster, they argued.[10]

Furthermore, on principle, ADB-funded projects were put up for international bidding: the lowest bidder gets the job. By contrast, Norad—in

---

[10] For a more detailed discussion of Melamchi as a single- or multi-use project, and the different national and international development institutions that advocated one side or the other, see Gyawali (2015: 12–15).

line with Hoftun's development philosophy—had made Nepali involvement a central priority, recognizing that real hydropower development in Nepal was about building up indigenous capacity, not just dotting foreign-built power plants across the landscape. For decades Norad had already devoted resources to support an independent hydropower industry in Nepal. Both sides agreed that there needed to be a tunnel but whereas ADB saw the infrastructure project as nothing more than a means to an end (to be contracted out to the lowest bidder), the Norwegians saw the project as an end in itself—to further strengthen Nepal's fledgling hydropower sector. It is difficult not to conclude (as Hoftun did) that only the second of these visions has anything to do with actual development: ADB's was a dependency-inducing handout, Norad's a capacity-building hand up.

Pradhan and Hoftun knew that time was crucial: the quicker matters could be pushed ahead the more likely things would go their way. They had momentum from official Norwegian and Nepali support, but a host of factors seemed to be dragging them to a halt. One was from within the Nepal government–controlled BPC board itself. Even though top officials had ordered BPC to take the lead on Melamchi, BPC's board-level leaders were anything but enthusiastic. Through the early months of 1998 Pradhan and Hoftun used their influence to try to get the BPC board to even *bring up* the topic of Melamchi but failed. Hoftun accused the government-appointed BPC board chairman of doing nothing more than push files around a desk and, by late March, Pradhan had concluded that "in spite of special instruction" straight from the minister of water resources, the BPC board will leave the Melamchi project "hanging" and "will never cooperate." Either because they were not interested in more work, or because they saw Melamchi as a step in the direction of privatization which they opposed, current BPC leadership was going to do all it could to obstruct. Hoftun and Pradhan tried to get board members replaced but in the end could only conclude that, as Hoftun said, "The privatization of BPC is a major factor in getting Melamchi moving fast." If Melamchi was going to happen, BPC had to be taken out of government hands.

Ironically, whereas the finance ministry's Privatization Cell should have been leading the way, if anything, they turned out to be an impediment to the privatization process. Rather than shepherding the process along speedily, Hoftun accused them of "dragging their feet" and delaying in order to make money. He blamed the British consultant for doing a "poor job" or at least "not being listened to" by anyone else. On April 1, 1998, Hoftun wrote again to Pradhan, "I think the privatization issue is now the most critical timewise" but neither were able to make any real headway. Six months earlier high officials had personally promised Hoftun and Pradhan that privatization would be completed *within the fiscal year* (mid-1998). When that time arrived, the Privatization Cell had not even managed to publicly release a "notice for tender." Already Hoftun fretted (presciently) that

> I am afraid that the potential investors in Norway and perhaps also in Nepal may lose interest because of the continuing slipping of deadlines. When HMGN delays and Ministers do not deliver what they promise, foreign investors lose confidence in Nepal as a country where they want to place their money.

One reason for the delay is that the Privatization Cell had decided that, before BPC could be put up for sale, the government needed to know the company's value. But figuring out BPC's worth turned out to be a huge point of contention. In what would be a recurring and fundamental failure of joint comprehension that would haunt the process for years to come, the government took the position that they were selling real estate, whereas Interkraft Nepal (IKN) understood themselves to be buying a business. For the government, BPC was a collection of power plants whose value existed in the present. But for the IKN investors, especially the Norwegians, BPC's current physical assets were secondary—almost negligible—compared to the company's *future potential*.

"Do they not realize that the value of BPC's property lies in the terms of contract for power sales?" asked an irate Hoftun when he learned of the Cell's intention to hire a Nepali "valuator." The power purchase agreements (PPAs) for both Andhi Khola and Jhimruk were set to expire in the near future. The terms set for those new power purchase agreements would have everything to do with BPC's future profitability but, not surprisingly, NEA refused to address the issue (knowing that it would aid the privatization process).

Furthermore, the IKN investors knew that BPC's ability to maintain and expand its power distribution and retail sales function was crucial to the company's future worth. Like the PPA, securing and expanding BPC's distribution area was a topic that *had to be* hammered out with the Nepal government in order for IKN to estimate the company's value and arrive at a bid price. "Do they not realize that this is something which in the end can only be decided through negotiations between the bidders and HMGN?" asked Hoftun. "A valuator cannot say anything about this." IKN wanted to buy a business with a future; the government wanted to sell a collection of real estate in the here and now. This disconnect is at the root of a failure that would plague the bidding process for years to come. From the IKN perspective, the entire bidding process was fundamentally flawed because the Nepal government refused to make clear the crucial contractual elements that would determine BPC's value.

Unable to make any progress with the government-controlled BPC board or the Privatization Cell, and worried that Norad money budgeted for tunneling already in 1998 would go to waste, Hoftun and Pradhan tried to jumpstart the Melamchi project through other avenues. Like at Jhimruk and Khimti, one of the government's commitments to the larger Melamchi project was to build roads to key construction sites. By early 1998 road construction was underway but advancing at such a glacial pace that it would be years before tunneling could begin even if all the other challenges were resolved. Rather than wait, Hoftun proposed building a ropeway, again as had been done in earlier BPC projects. In March 1998 Hoftun wrote to Pradhan that "the ropeway will be critical for the early completion of the Melamchi Project. It must therefore be started at the earliest possible time, without waiting for the total financing of the Melamchi Project to be competed first." Hoftun spent months courting a Norwegian ropeway manufacturer in hopes that they might supply and invest in a Nepal project. Both Pradhan and Hoftun worked hard to drum up investment support for the ropeway project which, they argued, would be a valuable piece of transport infrastructure both during and after the Melamchi construction phase.

In the meantime, apparently desperate to *do something* to keep the ball rolling, Hoftun tried to get tunnel construction underway. In May 1998 Dan

Jantzen (the American engineer who had earlier worked with the United Mission to Nepal [UMN] and BPC) joined Pradhan and Hoftun as part-time advisor to the Norad Melamchi project. Jantzen recalls that, when he began, "Odd's instructions to me were, 'Start the tunnel! Just start it. Go! We've got some money here.'" Several years earlier, as part of a feasibility study, crews had excavated about a half-kilometer test tunnel (to determine rock conditions) on the north side of the Shivapuri ridge along the "high tunnel" route favored by Norad that would include the hydel component. Jantzen tried to recommence tunneling but bureaucratic deadlock made any progress impossible. Norad money may have been budgeted for tunneling but until BPC privatization was completed, no money would be released.

Through the summer and fall of 1998 Balaram Pradhan kept up his grueling schedule promoting Melamchi and BPC's privatization. Records show that Pradhan spent virtually every day in a whirl of phone calls, office visits, home visits, and meetings with politicians and bureaucrats working every angle related to BPC, Melamchi, and privatization. Perhaps Pradhan's main challenge was the fact that the object of his lobbying was a moving target. With the government in turmoil and the cabinet's composition constantly shifting like a game of musical chairs, Pradhan had to repeatedly try to re-convince leaders how and why to move forward on BPC when they themselves were obsessed with the day-to-day politics of saving their own skins. Pradhan complained how, "at such political situation almost no one is interested to hear about or to comment about any subject aside from political intrigue!" Hearing of Pradhan's efforts, in August Hoftun commented, "It seems that as soon as the pressure by PM [prime minister], etc. is lessened, everything goes back to normal—nothing happens!"

Finally in early October 1998 Pradhan reported the breakthrough that the cabinet had "approved the proposal of selling 75% of HMG shares of BPC." The Privatization Cell proposed to accept bids for BPC by the end of December and finalize privatization by February 1999. But, warned Pradhan, the "Evaluation Committee" will be chaired by the secretary of MoWR whereas the "finalization of privatization" will be in the hands of the minister of finance. Because these ministries were controlled by different parties, there could be trouble. Nevertheless, Hoftun wrote enthusiastically to Pradhan, "I join your Minister friend in saying: Congratulations! Your efforts have been an important factor contributing to this Cabinet decision." The logjam seemed to have been broken, though already by November the Privatization Cell was a month behind schedule.

THE FIRST BID

In early November 1998 the Privatization Cell finally made the Butwal Power Company (BPC) bidding documents available to the public. Now the rush began to decipher the documents, analyze the bid conditions, finalize the bidding consortium's membership, prepare the formal bid, and outmaneuver potential competitors—all while trying to keep the government itself focused on the project and its responsibilities to keep things rolling. Balaram and Gyanendra Lal Pradhan worked hard to organize the final list of Nepali investors and determine how much that group would contribute to the bid. They also debated the prospect of joining forces with Indian hydropower developers but ultimately rejected that option in favor of keeping the project in Nepali hands.

Hoftun, with an eye to discouraging the competition who might not get the full story from the government, wrote, "Now it is time to prepare the notes on Power Plants which are negative and have not been mentioned in the [government's bidding] documents." Rather than put anything in writing, Pradhan instructed site managers at Andhi Khola and Jhimruk to make sure to inform any visiting potential bidders of problems like quartz sediment and erosion, water sharing, bad community relations, poor Nepal Electricity Authority (NEA) line quality, and so on.

By late December the Privatization Cell had sold eighteen sets of bidding documents to potential investors from Nepal but also firms from Australia, Hong Kong, Japan, Finland, the UK, and India. How many of these would actually submit a bid was anyone's guess and a cause for concern and speculation. Because one of the bid conditions was that investors had to have already developed and managed hydropower generation of at least 30 MW, none of the Nepali firms qualified on their own. But because the international firms needed Nepali partners, joint ventures became the order of the day as both sides sized up suitable partners.

A January 12, 1999, bidders' conference organized by the Privatization Cell attracted seven investor groups while also casting light onto some of the bid's political fault lines. Of the seven, Interkraft Nepal was by far the most knowledgeable and prepared. The only other group with international credentials was the Independent Power Corporation (IPC) (a British-registered firm with corporate roots in the USA), whose representative did not say a word. The others present represented five Nepali potential investor groups who took the meeting as an opportunity to air grievances. They complained that there was too little time between the bid's announcement and submission date only a few weeks away (January 31). They complained that it was difficult to get detailed information on BPC, its assets, and prospects. They complained that the price of the bid bond[11] was too high and needed to be reduced. And they complained that the bidding condition that purchasers must own and operate at least 30 MW of generating capacity should be waived as it disqualified any stand-alone Nepali bidders. Several groups threatened to lodge formal complaints against the Cell accusing them of "discouraging local bidders by putting clauses so that local bidders alone cannot take part."

But most tellingly, as Balaram Pradhan reported, the five Nepali investor groups complained that the entire bidding process was rigged in favor of Interkraft Nepal (IKN) and the Norwegians. "The government intends to give it to Interkraft," said one representative. "The ministry of finance is set to give BPC to Norwegians," complained another. In fact, this was a point of contention that had surfaced before and would come up again repeatedly in the years to come. Even reasonable people could suspect a kind of arm-twisting policy on the part of the Norwegians. On the one hand, Norad had made its Melamchi money contingent on hiring BPC (as a *Nepali* company) to carry out an important part of the project. On the other hand, the same Norad grant demanded BPC's privatization—a process that favored rich foreign bidders and, if IKN was successful, would result in a "Nepali"

---

[11] A bid bond is a sum of money that bidders are required to deposit in a special account to guarantee the bidder's ability and willingness to carry out the terms of the bid transaction. Bid bonds have to be maintained throughout the bidding process and only revert back to the bidders once the sale is finalized.

company with Norwegian majority owners. The implication seemed to be that if Nepal did not "give BPC to the Norwegians," Norway might withdraw its support for Melamchi.

Balaram Pradhan was especially sensitive to this criticism (of IKN enjoying insider privileges) because he had to deal with it on a regular basis. After a meeting with government officials Pradhan reported to Hoftun, "One thing I understood from them was that HMG has noted privatizing BPC means selling of shares to Norwegians, and I have tried to convince them that it is not exactly true." "Not exactly true" is not exactly a denial. At another point Pradhan warned Hoftun that, in order to "avoid unnecessary scandal," IKN had to discourage any perception that Norway was trying to extort BPC out of the hands of the Nepal government. Hoftun himself was ambivalent on the point. On the one hand he believed IKN should win the bid on the basis of its superior qualifications. But on the other he felt that Norwegians had every right to lay claim to BPC given that country's decades of state and commercial support of the company. If IKN had an unfair advantage it was not because of Norwegian extortion or, at least, not exactly.

Even if it was only an illusion, many people inside and outside the government sensed that it was real or at least used the apparent conflict of interest as ammunition to fight the privatization process. Inside the government, people who wanted to hang onto what they felt was a valuable government asset (or, perhaps, source of illicit income) argued that BPC's privatization was a Norwegian robbery plot. In the private sector, especially among those interested in bidding for BPC, it was easy to argue that the Norway and Nepal governments had colluded to stack the deck in IKN's favor.

Unfair or not, the weeks following the bidders' conference were a hectic blur for IKN representatives trying to prepare a comprehensive, detailed, and carefully presented bid. While Pradhan represented IKN in Kathmandu, in Norway Hoftun, Jantzen, Ratna Sansar Shrestha (as finance and legal expert), and Norwegian Interkraft representatives worked through successive drafts of bid documents.

The Privatization Cell requested that the bid proceed through a "two envelop system" involving separate technical and financial bids. Bid evaluators would score the quality of technical bids numerically out of a possible 100 points. Only those scoring 75 points or more would have their financial bids opened.

- 15 points: bidder's capability and credibility
- 40 points: bidder's proven experience in hydropower development, including in countries like Nepal
- 30 points: business plan including proposed investment, marketing, training, employment, technology transfer, environmental concerns, and so on
- 15 points: broad impact on Nepal's economy and involvement of Nepalis in plans

IKN's bid included detailed discussions of every point above, almost all of which bear the distinct stamp of Hoftun's development philosophy, perhaps best summed up in the following:

> Most fundamentally, the consortium does not seek to dominate and replace the Nepalese capability that has already been built up within BPC, but rather to support and expand this capability, and to help it grow and achieve maturity. The Bidders expect a reasonable return on its investment, but otherwise plan to reinvest additional profits in the further expansion of BPC. By demonstrating how a well-run, efficient, and honest company can benefit its customers and still achieve a reasonable return on investment, the consortium hoped to influence not only the electric power sector of the Nepal economy, but other public and private enterprises as well.[12]

IKN felt confident that, with its combination of experienced Norwegian utilities and the experience and insights of Nepal specialists like Hoftun and Pradhan who had actually guided BPC for decades, they would easily outscore any others in the technical bid.

---

[12] From the "Executive Summary" of the technical bid, dated January 31, 1999, 22.

The second "envelope" or financial bid—where bidders named an actual figure for what they were willing to pay per share for the shares offered— proved to be much more difficult and controversial. As already mentioned, IKN understood BPC's value to be a function of what the government would be willing to pay for power generated in the future, and of what deal BPC could work out with the government for private control of power distribution and retail networks. For many months Balaram Pradhan had pushed officials and bureaucrats in Kathmandu to provide: (1) if not a new formal power purchase agreement (PPA) for BPC's power plants, then at least some kind of guarantee for the future, and (2) official guarantees regarding the private distribution area that BPC could count on managing in the future.[13] Pradhan warned that the government *had to* address these concerns *before* IKN could submit its bid. But ministry of finance and Privatization Cell officials brushed off IKN's concerns saying dismissively that, "bidders will have to take the risk." Again, the government thought of itself as selling inert infrastructure "as is," whereas IKN saw BPC as a dynamic entity whose long-term value depended on local contingencies that had to be taken into account.

Faced with government refusal to address its concerns, IKN submitted a complicated bid that involved one "main bid" and two "alternative bids." IKN's main bid was for NPR 130 per share (totaling c. USD 11.6 million) but, crucially, that figure was *conditional* on a PPA being reached with specific values for power purchase from Andhi Khola and Jhimruk: NPR 3 per kilowatt hour for wet season and NPR 4.25 per kilowatt hour for dry season generation. Thus IKN's main bid was high but conditional.

---

[13] It is important to note that the idea of a private distribution area for BPC was neither new nor its own. In fact, MoWR had previously contacted BPC on several occasions to discuss the possibility of its taking over responsibility for power distribution in a ten-district area of west Nepal. The BPC board had already approved the government's request. What's more, the government's own official 9th five-year plan (1998–2003) included the "unbundling of NEA" as one of its official objectives. That is, at least at the official ideological level, the government wanted to break apart NEA's power monopoly by opening up the market to private entities. But here, as in many other cases, the government's officially stated wishes had little relevance for how individuals and agencies carried out the government's actual operations.

IKN's first "alternative bid" was for the entire BPC package in an "as is" condition. With no PPA or distribution area guarantee, IKN would pay only NPR 90 per share. In the bid IKN again reiterated that they had no way of judging BPC's value without this information. In fact, they did not even have an official guarantee that the government would continue to purchase power from BPC *at all*. The second alternative bid had no specific price attached to it but consisted of a proposal to negotiate a share price based on the size of a distribution area the government might hand over for BPC to expand its operations into. If the government was looking for a straightforward, unambiguous bid, they did not get it from IKN.

During several weeks of intense scrambling in January 1999, IKN participants phoned, faxed, and emailed documents and comments around the world in Norwegian, English, and some in Nepali in order to wrap things up in time for the bidding submission date of Sunday, January 31, 1999.

Then disaster struck: the bottom fell out of the IKN bid. The first inkling came on Tuesday, January 26, when Hoftun reported that "the Trondheim utility," one of the five IKN Norwegian investors, had backed out of the BPC venture. Hoftun called this "sad news" that has "weakened the financial base" but, overall, did not seem too alarmed. The real calamity hit on Friday the 29th. Writing some months later, IKN Board Director Gunne Hegglid described how a "sunshine" period had "lasted until the miserable Friday, 29 January, a day which at least I will never forget." Even as their representatives sat in Kathmandu preparing the final documents, on that day word arrived that the remaining four Norwegian utilities were withdrawing from the bid. Only one of them, SKK Energi, agreed to stay involved but now only in a (paid) advisory role. Two days before the bid was due, IKN had lost almost 90 percent of its bidding funds.

Ironically, this turmoil in the effort to privatize BPC was the result of similar forces at work in Norway. Not unlike the transformations that swept through Statkraft (described in the previous chapter) as that Norwegian government-owned hydropower developer underwent market-based commercial restructuring, the late 1990s saw similar pressures exerted on Norwegian publicly owned power utilities. Even as the Norwegian Interkraft power consortium was negotiating its bid for BPC, market and political forces back in Norway suddenly tore it apart, thereby pulling the plug on BPC investment.[14] "The confederation had broken up, back here in Norway!" recalled Hoftun in 2015. "So I can't blame the Nepal side for all of this. It was in a period here in Norway of great upheaval. The whole energy business was more or less privatized too—taken away from the government and commercialized."

In a frantic, nightmarish twenty-four hours, IKN refinanced its bid. Hoftun had about USD 1 million at his disposal from a Norwegian mission-based investment fund. For the rest he turned to the Nepali investor group. Would they be willing to go to the banks and get loan guarantees for about USD 10 million to go toward the bid? Hoftun and Hegglid (the IKN board director) promised that they would immediately go to work recruiting other Norwegian investors with the full intention of purchasing a controlling 51 percent interest in BPC. But that would take time. Could the Nepalis float the bid until that time?

For the Nepali investors the entire situation came as an enormous shock. As one of the leading members of the Shangri-La Energy Ltd. (SEL) group told me, up to that point few in their group had given the matter of BPC ownership a lot of focused attention. As individual firms their investments were relatively limited and, with the Norwegian utility investors taking the lead, the Nepalis could, more or less, sit back and enjoy the ride. Now, suddenly, the Nepalis were asked to guide the entire ship even if (in theory) only temporarily. Would they expose themselves to the now very real risks involved in becoming the lead investors in BPC without active substantial Norwegian involvement?

At this point, as the SEL member I spoke with acknowledged, the strictly business aspects of the deal became clouded with strong emotions. The IKN bid was about to collapse. The IKN bidders had already known for some time that at least one other competitive bidder, the British IPC, was in the game and were rumored to have teamed up with the Nepali Chaudhary Group. "Are we going to let Marwaris take over BPC?" the IKN Nepali investors asked as passions began to rise. Dan Jantzen, who was in the thick of things

---

[14] At the point of its breakup, the Interkraft consortium sold the "Interkraft" name for a nominal price to the Norwegian mission organization with which Hoftun was associated after his retirement from the United Mission to Nepal (UMN).

in Kathmandu, remembered how "by this time there was intense nationalism around the whole thing. 'By god we're going to beat them!' and like this. The juices are starting to flow. And here we are with no bidder!"

* * *

For readers unfamiliar with Nepali cultural politics, a word about Nepal's Marwari community may be in order. Originally from the Marwar district of Rajasthan in northwestern India, Marwaris are members of a prosperous Hindu merchant caste that, for centuries, have formed family-based trade networks across South Asia and beyond. Today their ranks include some of the wealthiest people in South Asia including the famous Birla family of industrialists. In Nepal, Marwari families have been resident since at least the Rana era and, in many cases, have been Nepali citizens for generations and include many successful businessmen (Chaudhary 2015). Nevertheless, in the eyes of most Nepalis, Marwaris remain "Indian" regardless of the depth of their ties to Nepal. Though in many ways model citizens, the Marwari community draws suspicion not only for its wealth and Indian business connections, but because of its extremely insular, culturally conservative, closed nature that—at least until recently—included the isolation of its women and marriage alliances with Marwari families in India. Combining envy and suspicion with competitiveness, many Nepalis believe that Marwaris do not have Nepal's best interests at heart.

This is certainly true of popular attitudes toward the Chaudhary Group and its owner Binod Chaudhary, the richest man in Nepal. In 2017 *Forbes* magazine estimated Binod Chaudhary's net worth at USD 1.33 billion, making him the world's 1,563rd richest person and Nepal's only billionaire. Odd Hoftun, in an interview with me, dismissed the Chaudhary Group as representing nothing but "Indian capital," viewing Binod Chaudhary as a virtual agent of Indian colonialism. He defended IKN's crusade against the Chaudhary Group as a way to "contribute to Nepal's ability to stand on its own and not just be a colony of India…. So [we took them on] both to carry out a process that would be clean and prevent BPC from ending up in Indian hands." Of course Binod Chaudhary is not an Indian but for Hoftun—who many Nepalis described as more of a Nepali nationalist than them, and who had certainly absorbed all of the common Nepali prejudices—Marwaris would always be Indians. Under Chaudhary Group joint ownership, BPC would not have suddenly become Indian—though it is worth noting that years later when Binod Chaudhary successfully purchased Himal Hydro, he turned it into a subsidiary of an Indian construction company. For Hoftun Binod Chaudhary's sin was of being more loyal to capital (Indian or otherwise) than to Nepal.

* * *

With Nepali nationalism coursing through their veins, the SEL investors went to their bankers, secured USD 10 million in loans, and saved the bid. At this point none of the Nepali investors wanted to shoulder the majority of the bid for long but were willing to step in to allow the Norwegians time to find new foreign investors who would again commit funds for a controlling interest. (Ironically the Nepali investors were comfortable with Norwegian control of BPC, but not Marwari.) "It all happened in no time at all," remembered one participant. IKN deposited its bid documents with the Privatization Cell, cover letters dated January 31, 1999.

THE COLLAPSE OF THE FIRST BID AND THE END OF BPC INVOLVEMENT IN MELAMCHI

Unfortunately, what the bidders thought would be more or less the end of an exhausting process only turned out to be something like the beginning. According to the bid documents, the bid evaluation process was to take about a week and, in fact, the Privatization Cell opened the technical bids almost immediately. Only one other party, the British Independent Power Corporation (IPC), had submitted a bid.

Then everything stalled. In February 1999 the sitting government dissolved resulting in new cabinet appointments and new members in the Privatization Cell. Pradhan reported dryly that "these days the government departments don't seem very interested in any other business than politics." By early March Pradhan learned that the bid documents had not been looked at since their opening. When he reported this to the minister of water resources, the minister "could not believe it" saying that he and the minister of finance favored privatization of the Butwal Power Company (BPC) and wanted to see it through quickly. Pradhan had to inform the minister that his own secretary was supposed to have formed a bid "evaluation committee" but had not. Later Pradhan reported that this secretary, a political appointee, knew virtually nothing about power generation ("he is not a technical man") or BPC. "I think he needs lots of information to be explained in a simple way," said Pradhan. As of late March, nothing had happened and there appeared to be "little hope that this committee will start working soon."

The upside to all this delay was that it gave Hoftun and the other Norwegian Interkraft Nepal (IKN) board members time to recruit new investors for the BPC bid. By late April Hoftun reported that they had raised around USD 4 million worth of promised funds from various sources including two of the Norwegian utilities that had now agreed to make relatively small (for them) investments. With this money, along with what he hoped would be matching funds from a new Norwegian government investment program (Norfund), Hoftun was again optimistic. "It seems that the Norwegian group are now definitely ready for 51%," he reported to the Nepali investors.

By this time Pradhan had news—good and bad—from the Privatization Cell. The Cell's British advisor, Douglas Clarke, told Pradhan that an evaluation committee had indeed been appointed and had begun work but progress was very slow. The committee met only sporadically and, because the documents they needed to consult were locked up except during meeting times, even the meetings were inefficient. Worse still, Clarke reported that the new Cell members were all ministry of water resources appointees and Nepal Electricity Authority (NEA) sympathizers opposed to BPC's privatization. In the hope of scuttling the bid, "they are trying to find a good reason so that they could fail one of the party and propose for retender saying that only one bidder cannot be negotiated," Pradhan reported. At the end of May 1999 the evaluation committee finally submitted its report to the concerned ministries but repeated cabinet reshuffles meant that the privatization process languished.

After almost half a year of waiting, June 1999 saw a flurry of BPC-related developments. Early in the month the finance minister announced that the financial bids would be opened in a week and that the handover to new owners should be complete by mid-July. No one expected this to happen and, in fact, the bid opening did not occur until June 18.

It was at this point that the whole BPC privatization process veered sharply out of the government shadows and into the light of public debate and opinion. When bids were publicly read at the opening ceremony, Cell officials only announced IKN's "as is" unconditional bid of NPR 90 per share, not what it felt was its "primary bid" of NPR 130 per share. (Cell officials announced that conditional bids were forbidden in the bid documents, a point strongly contested by IKN and their reading of the ambiguous language.) By contrast, IPC's bid was, at NPR 109.09, almost NPR 20 per share higher. Worse still, IKN learned that its technical bid score was only 76 (barely past the minimum) whereas IPC's was 92 (Chaudhary 2015: 259), an almost inconceivable outcome from IKN's perspective. Finally, the opening confirmed IKN's fears and suspicions that IPC's local partner was indeed the Chaudhary Group.

Suddenly the IKN group found itself broadsided by a barrage of negative press coverage orchestrated by the Chaudhary Group. Apparently in direct violation of the confidentiality clauses contained in the bid documents, the Chaudhary Group held press conferences and issued press statements in a direct attempt to sway the government's decision regarding BPC by swaying public opinion. Chaudhary boasted of their higher technical score and complained that IKN had violated bidding terms by submitting a conditional bid.

After having its formal complaints to the Privatization Cell thoroughly ignored, IKN decided that—if the other side was ignoring protocols—it was time to take off the gloves and play by the other side's rules. In a series of strategy meetings the IKN side debated "how to recover and retake the initiative" from IPC and the Chaudhary Group. The IKN investors felt profoundly cheated. IKN's main financial bid had been higher. And somehow IPC had managed to get a higher "technical evaluation"—"perhaps because of payoffs," IKN speculated darkly.

Hoftun felt that the facts favored IKN. In a memo from late June Hoftun summed up his comparison of the main bidders in two points: (1) IKN represents utilities with close to a century of experience developing, operating, and distributing hydropower while IPC is a relatively small investment company with little experience in operation and engineering, and no experience in building or operating hydropower plants; and (2) IKN has, directly and indirectly, decades of experience in Nepal through various forms of Norwegian involvement in developing the hydropower sector. This is reflected in their detailed and specific business plan. By contrast, IPC has virtually no experience in Nepal and its business plan amounts to little more than promising to invest large sums of money. As IKN set out to sway opinions, Hoftun concluded that "these facts give the Norwegian/Nepali bidders a strong position when facing HMGN both in informal lobbying and in formal negotiations."

But having the "facts" on IKN's side did not do much to calm Hoftun's nerves. If anything, official news of the Chaudhary Group's involvement only gave vent to all of Hoftun's worst fears about the power of corruption in Nepal. For Hoftun the Chaudhary Group was synonymous with the "cancer of corruption" that "through all my years in Nepal I have resisted." Hoftun warned the IKN investors that Chaudhary Group involvement "must not ... scare us into the trap of 'under-the-table' dealings." He even went so far as to send a "personal and confidential" letter to a high IPC official warning them of the Chaudhary Group's reputation for corruption. "You are surely aware of the problem of corruption in Nepal," Hoftun wrote. "But you may not know that your local agent[15] as well as your Nepalese partner have a dubious reputation in this respect." He goes on to say that "my contacts" now fear that "the privatization of BPC will be decided by who pays the most under the table. That is why I am writing to you now.... I am sure that your own Corporation would not want to engage in any illegal practices in Nepal. But it is very easy to get trapped into this through local agents and partners."

Hoftun was not the only one on the IKN side who feared Chaudhary Group machinations. The Nepali investor group, and especially their leader Gyanendra Lal Pradhan, were sure that the other side was up to "dirty games," busily buying their ways into the hearts of key government officials. Some

---

[15] IPC's "agent" was Inter Continental Trading Concern, Pvt. Ltd., a Nepali firm specializing in providing liaison services for foreign investors looking to do business in Nepal.

argued for "the necessity of under-table business" due to "a strong feeling that we will lose otherwise." But Hoftun and Pradhan held firm arguing that the Privatization Cell, with its international standards and foreign advisors or watchdogs, would guarantee and reward "clean business practices."

IKN would not go "under the table" but it would jump into the fray in the court of public opinion. IKN launched a publicity campaign to explain and promote their side, including how the bid documents did not, in fact, prohibit conditional bids. They agreed to redouble their efforts to push the government to negotiate a new power purchase agreement (PPA) for BPC's power plants. And they even eventually agreed (despite strong misgivings among the Norwegian board members) to drop the conditions on their "primary bid," thereby raising their "as is" offering to NPR 130 per share and putting them well ahead of IPC and the Chaudhary Group's bid. This move caused further howls from the IPC side who claimed that IKN was violating bidding terms by changing their main bid.

Perhaps IKN's boldest move was its attempt to overtly rebrand their bid in the eyes of the public by linking it to the Melamchi water project. In both the press and in their lobbying, the IKN side agreed to promote the BPC–Melamchi connection and underscore the importance of IKN and the Norwegians getting the bid in order for Melamchi to progress quickly and smoothly. Earlier IKN had hoped that its "10 District Concept"—in which an IKN-controlled BPC would take over distribution and retail in ten western districts—would be a powerful selling point. But with neither the government nor the public showing much interest in the idea, IKN decided it was time to hitch the BPC privatization wagon to the Melamchi project which had much more enthusiastic public and official support.

For the previous six months, debates over BPC's role in the larger Melamchi project had been inconclusive: would the water delivery system include a hydropower-generating component, or not? Norwegian aid money said yes; the Asian Development Bank (ADB) said no. But Pradhan and Hoftun could feel the momentum they had enjoyed initially slipping away. Some argued that a lower tunnel (without the drop needed for power generation) would be slightly shorter and therefore cheaper and quicker to build. By February 1999 rumors circulated that the World Bank, which was funding other parts of the larger project, opposed the tunnel altogether on the grounds that it was too expensive. They favored pumping water into the Kathmandu Valley. Norad officials were still holding their ground, assuring Hoftun that, in their opinion, other ideas were "sheer nonsense" and that "nobody who understands the situation" favors removing the power component, including Nepal's prime minister. But with no real progress being made on any front, by June 1999 it was time to make a new aggressive sales pitch, this time linking Melamchi with BPC's privatization.

Dan Jantzen took the lead in promoting what came to be known as the "BPC for Melamchi" package. In a "concept paper" entitled "Combining the Issues of BPC Purchase and Melamchi Development" dated June 27, 1999, Jantzen argued that IKN was making little headway using other tactics and therefore it was time to "tie our bid for BPC to the development and construction of the Melamchi Diversion Scheme.... Put simply, the consortium could offer to build the Melamchi Diversion Scheme in exchange for [the government's] 75% of the BPC shares." In return for handing over its BPC shares to IKN, IKN would use already-committed grant money, as well as mobilizing sizeable private investment, to build a tunnel and commercial hydropower plant, the discharge from which would be a free water supply for the Kathmandu Valley. IKN would get BPC and the Melamchi development contract, and the government would get free water.

In the time since Jantzen had joined the BPC privatization effort he had put a lot of new thought into the Melamchi project and these new ideas first appear in his June concept paper. In what later became known as the "Multi-Purpose Melamchi Project" (MPMP) Jantzen proposed a considerably expanded scheme that would tap into not just one river, the Melamchi Khola, but a total of four, also including the Yangri, Larkhe, and Balephi Rivers that run above 1,700 meters north of Kathmandu.[16] Jantzen pointed out that

---

[16] According to Jantzen's concept paper, the proposed "BPC for Melamchi" swap would require the government's transfer of 75 percent of BPC's shares to IKN but also the following: providing the USD 24 million Norad grant plus the promised USD 6 million Nepal government grant to the consortium; providing an investment-grade PPA for Melamchi, Jhimruk, and Andhi Khola power; making land for Melamchi construction available to BPC; building the necessary access roads for project construction; providing facilities specified in various acts—duty free import, tax relief, and so on; providing water rights to the Melamchi, Yangri, Larkhe, and Balephi Rivers

(as very few people in Kathmandu realize even today) while Melamchi River water would be a welcome addition to the valley's water supply, it would not come near to quenching current, let alone future, demand—especially in the dry season. However, with additional tunneling, it would be possible to divert much more water guaranteeing a year-round steady high-volume flow.

Increasing flow through the already proposed tunnel would mean more drinking water for Kathmandu but it also had other very attractive implications. First, it would mean more power-generating potential for the already-planned power plant: more water volume, more megawatts, more power for Nepal, more profit, more appealing to investors. Second, it would mean the return of a clean, free-flowing Bagmati River that would have both aesthetic and health benefits for the Kathmandu Valley. Third, with its higher year-round flow, more power plants (with hundreds of megawatts of potential) could be built along the Bagmati *below* Kathmandu as it flowed to the Tarai.[17] And finally, all of this water would flow directly into large Tarai irrigation schemes that were already built but chronically short of water, especially during the dry season. Jantzen and IKN argued that, though greatly expanded, this plan also promised enormous benefits and, best of all, all or most of it could be funded privately through developing the plan's hydropower components commercially (Jantzen et al. 2008). Furthermore, it could be built piece by piece, with the original Melamchi tunnel its first priority. IKN offered to deliver free water to Kathmandu by 2005.

"BPC for Melamchi" was a bold scheme including some visionary ideas, but it was also an attempted end run around the entire international competitive bidding process that the government was ostensibly committed to. With enough forceful backing the government could have pushed through such a deal but there were also powerful forces aligned against it. IPC and the Chaudhary Group called foul, accusing IKN of trying to short-circuit the bidding process. They also accused Norad of trying to sway the process in favor

---

above 1,700 meters; and providing preferential rights to construction of the Yangri, Larkhe, and Balephi tunnels.

[17] The MPMP proposed a series of three hydropower plants—one at the outlet of the originally planned Melamchi tunnel and two along the Bagmati River below the Kathmandu Valley—with a combined total of 650 MW.

of the Norwegians by implicitly linking grant funding to IKN's successful bid for BPC, a line soon picked up by the press. Interests within the government, and especially within NEA, opposed privatization altogether: why privatize a government asset that was bringing in around NPR 1 billion per year in revenue? And the major international funding entities, who already saw even the most basic tunneling and hydropower elements as too complicated, rejected the expanded project as so much pie in the sky. They also questioned IKN's capacity (and capital reserves) to carry out even the simplest version of the scheme.

Through the summer of 1999 both of the main bidders, IPC and IKN, upped their lobbying games and fed information to the press backing their bids. By late July affairs had become so heated that Balaram Pradhan feared that his IKN email account was being hacked by Chaudhary Group agents! But while there was plenty of smoke, there appeared to be very little fire. The Nepal government again seemed to be at a standstill, immobilized by its own internal chaos and the rising threats emanating from the People's War. In August Pradhan sounded discouraged as he reported on round after round of government briefings that seemed to have no effect. "One has to have a lot of patience to keep on pushing again and again," he lamented.

Worse still, Pradhan reported growing pushback from government officials. After stressing the importance of a new PPA, one official complained that Nepal had been "cheated in Khimti" (by a rate too favorable to the developers), prompting Pradhan to note that "he does not have a very soft corner for Norwegian companies." Furthermore, the official blamed it all on "Hoftun [who] has been the planner" of the whole thing. Pradhan again explained that the Khimti PPA was not Hoftun's fault but that of the "development" banks. But still, the anti-Norwegian, even anti-Hoftun, sentiment was disturbing.

Seeing no progress, the British and American embassies began pressuring the ministry of finance to conclude the BPC privatization process, with the British even threatening to withdraw support for the Privatization Cell if they did not see quick progress. Sensing foul play, Hoftun condemned the two embassies for "shamelessly lending support to the IPC bid" in order that a British–American company could gain a foothold in Nepal. Even more frustrating for Hoftun and Pradhan was the fact that the Norwegian embassy, for its part, refused to publicly endorse the IKN bid. In fact, if anything, the Norwegian embassy's Norad representative seemed to be gradually backing away from the Melamchi project's hydropower component, and BPC along with it. For months Pradhan had met privately with this official but by late in the year even she had concluded that the power component added too much cost and time to the larger project and that there were better, less costly places to build the same size power plant as the one IKN proposed.

Perhaps the final nail in the coffin for IKN and BPC's hopes for a Melamchi hydropower project was the matter of rising costs. In late 1999 Statkraft Anlegg, the Norwegian construction company (and its partner Himal Hydro, then wrapping up work at Khimti), released a new estimate for the tunnel and hydel component which put the total cost at USD 162 million, 50 percent more than what Hoftun and IKN had estimated. Hoftun wrote to others in the IKN circle that he was "deeply shocked" at the news. Statkraft's estimate increasingly made the entire project untenable, and even impossible following the "Jhimruk model" of cost containment. Even before this point there had been talk of altering the "BPC for Melamchi" pitch (with its promise of free water), acknowledging that a water tariff might be necessary. But this estimate was a devastating blow. "As far as IKN is concerned," wrote Hoftun, "it is almost like the sudden financial collapse just ahead of the bidding deadline of 31 January [1999]. I simply cannot understand what has happened." But he had suspicions. Later Hoftun accused Statkraft of inflating their estimate in order to protect themselves from unrealistic or imagined risks and, generally, to give themselves bargaining room if and when they entered negotiations for a lower price. But in the meantime, the Statkraft estimate only seemed to confirm what anti-hydel component critics had long been saying.

THE SECOND BID

Finally in late September 1999 word came that the ministry of finance was requesting "clean bids" from both IKN and IPC by October 6 in which their "Offer and Payment terms" were to be "clearly defined." Unhappy with the multiple bids, assorted irregularities, delays, and public controversy surrounding the first bids (the blame for which the government disingenuously laid at the feet of the bidders), the ministry called for new, unambiguous bids for BPC in an "as is" condition.

In desperation, Pradhan and Jantzen went to the Privatization Cell to beg for a new PPA, or even just a statement of intent to renegotiate a PPA, before IKN was forced to submit a new bid. How could the government be so insensitive to the legitimate concerns of the bidders, they asked. But the only person they could find to talk with, the British advisor Douglas Clarke, told them apologetically that the members of the Privatization Cell were political appointees with no understanding of business whatsoever. What's more, the government (including NEA, the agency that would have to negotiate a PPA) was divided and virtually paralyzed with almost no agreement between, and often even *within*, its various ministries. Rather than hope for a new PPA, Clarke advised IKN to base their bid price on the "good will value" of BPC and its positive reputation in Nepal. Pradhan had to explain that neither the Norwegians nor Nepalis in IKN were willing to "put down a single penny for good will."

The decision of where to set the new bid price fell to the primary Norwegian investors and their representatives on the IKN board—the people with the most money at stake. Correspondence and memos suggest that these board members were already upset at what they felt as having been forced to formally drop the conditions on their original NPR 130 per share bid. Hoftun, Pradhan, and the other Nepali IKN investors felt that even that price was too low. Understanding Nepali history, politics, and economics, they understood BPC's value and potential. But from the Norwegian point of view, the optics of 1999 Nepal were profoundly discouraging. What from a Nepal perspective looked like business as usual or, at worst, a temporary rough patch, for the Norwegians looked like a state and society on the verge of collapse. For Nepalis, investing in BPC looked like a good business opportunity whereas for Norwegians—accustomed to a stable and predictable political and business environment—investing in BPC increasingly looked like investing in the Titanic adrift in a sea of icebergs.

In spite of "strong doubt" and "feeling strong pressure from our Nepali friends," the IKN board set the second bid price at NPR 115 per share. Knowing that the price would disappoint the Nepali investors (because it was NPR 15 per share *lower* than their earlier bid), the IKN board director apologized but said flatly, "IPC can beat us, but IKN has stretched as long as possible." The IKN board director also wrote a lengthy memo[18] to be included in the rebidding papers that scathingly criticized the Nepal government for failing to meet deadlines and the Privatization Cell for incompetently handling the process and, in a very thinly veiled reference, accusing it of corruption.[19] Ahead of the October 6 rebidding deadline Balaram Pradhan reported "feeling that it is a suicide case." Gyanendra Lal Pradhan called for a bid price of at least NPR 135 per share arguing that any bids below that would give the government justification to simply cancel everything and retender the bid.

Yet, to everyone's surprise, when the documents were opened on October 6, 1999, IKN had outbid the IPC/Chaudhary Group consortium. IKN had lowered their bid to 115 but IPC's bid had remained unchanged at 109.09. In fact, in the days leading up to the rebid there were rumors that IPC was getting cold feet and might not bid at all. IPC's bid terms suggested that they were backing away for fear of the risks involved. The IPC bid of roughly USD 10 million offered only 1 million in cash at closing with the other 9 coming later from loans against BPC assets. In other words, IPC was willing to sink almost none of its own funds into the BPC purchase. Having expected disaster, the Nepal side of the IKN consortium left the second bid opening hopeful that things were finally going their way.

At this point, rather than wait for the government to soberly evaluate the bids, both IKN and IPC redoubled their efforts to lobby officials and influence public opinion. The bidders' bid bonds were set to expire on November 25 so both sides assumed that a decision would be reached by then. Unfortunately the last months of 1999 were a time of exceptional political uproar even for Nepal. The Nepal Congress strongman Girija Prasad Koirala

---

[18] Entitled "MEMO on the Situation for the BPC Rebidding Process on 6 Oct. 1999 as Seen from Interkraft Nepal AS," signed by Gunne Hegglid, director of the IKN board.

[19] Hegglid complained that, while IKN had tried to follow a "clean path," various opponents, including the government, had placed "stones" in their way trying to "force" IKN to "break our principles." He wonders whether it was IKN's "attitude to avoid corruption" and, by implication, the Cell's unwillingness to clear those obstacles, that was responsible for the long delays.

had been forced to hand the prime minister's office to his Congress rival Krishna Prasad Bhattarai, but Koirala remained the real power broker even as both men struggled for control of various ministries which were, therefore, often in turmoil and at odds with each other. Though fertile ground for intrigue, it was much less conducive to getting complicated decisions made that necessitated cooperation across several ministries.

Into this twisting political landscape IKN sent a team of six people— Pradhan, Hoftun, and Jantzen but also Bikash Pandey, Ratna Sansar Shrestha, and K. B. Rokaya[20]—to try to promote IKN's bid for BPC but also launch a last-ditch effort to sell the Melamchi project *with a hydel component.*

Once again the links between Norwegian support for Melamchi and the question of BPC privatization came to the fore. IKN wanted to avoid the impression that Norway was pressuring the Nepal government by making its commitment to Melamchi conditional on BPC going to IKN. But on the other hand, Pradhan, Hoftun, and others desperately wanted Norwegian officials to at least go to bat, publicly or privately, for the IKN side. In an August 21 email, Hoftun discussed strategy with Pradhan, noting that while IKN cannot be seen to imply that conditional demands are at play, they can make the case for Norwegian control indirectly. Given that Norad wants the Melamchi project to benefit Nepali companies, then "by simple logic, Norad wants IKN to be the new owners of BPC," Hoftun wrote. "It goes without saying that Norad does not want to see its thirty-year-long effort to build up BPC as a Nepali utility go to waste by being gobbled up by international big capital." (Hoftun saw IPC as nothing more than a corporate bloodsucker that aimed to drain BPC, and Nepal, of its money.) Yes, Hoftun conceded, BPC *should be* in Nepali hands. But given that the government had set the bidding conditions such that only foreign parties could meet its qualifications, then foreigners were going to control BPC. Certainly better that those foreign owners be concerned Norwegians with deep ties to BPC and Nepal than immoral "international big capital" (IPC).

By October 1999 the Norwegian members of the IKN board were reaching the limits of their patience. At an IKN board meeting late in the month the

---

[20] A prominent public figure, Christian, and human rights champion with some engineering background.

group decided, against Hoftun's wishes, not to renew its BPC bid bond past its November 25 expiration date. Reporting on the group's decision, Hoftun wrote, "IKN's bid is final as given, there is nothing more to discuss.... So now it is just for HMGN to make its decision." The remaining industrial investors have "lost faith in HMGNs ability or will to handle the bidding procedure in a proper manner." They are "tired and fed up and ready to pull out." If the government is disappointed at the low bid prices, "they can only blame themselves for having spoiled the whole thing and brought down the value of BPC in the eyes of bidders by these endless delays and their inept way of handling the BPC privatization process from the beginning to the end." "If HMGN does not accept the price offered, then this chapter is closed."

Hoftun tried to convince the board to renew the bid bond to let the process work itself out, but they refused. He tried arguing that "Nepal is different from Norway, etc. But in the long run it does not work. They are losing their trust in me as well!" Again on November 21 Hoftun made an appeal to the board to renew the bond, noting developments that appeared to suggest a possible favorable outcome for IKN. But on November 23 came another "no." "It is really shocking news for our Nepali friends," said Pradhan. "I don't know how they will receive this." Hoftun too was "thoroughly disappointed" seeing the move as another Norwegian betrayal like the one on January 31 (1999) when the Norwegian investors pulled out.

> Each time it has been the Norwegian parties I have tried so hard to bring together, who have failed.... Personally I am angry about the manner this has taken place, having been pushed out of the process.... I am ashamed to face the Nepali partners after this. I don't know how to explain what has happened.

The next day Pradhan broke the news to government officials who were sincerely surprised and perplexed that the Norwegians had not renewed—despite IKN's repeated promises *not to renew* (if negotiators delayed unnecessarily)! "I told them flatly that Foreign Investors cannot follow the way we work. So if we want Investors to come we have to work the way they get attracted." It seemed to be a genuine case of culture clash: for the Norwegians "Time is money" but the Nepalis wondered, "What's the rush?"

With the Norwegians having washed their hands of the business, Hoftun was already thinking ahead to the next retendering.

Yet the bid refused to die. Somehow the Nepali investors and Hoftun persuaded the IKN board to renew the BPC bid bond (on December 8), almost two weeks after it had lapsed. The decision was part of a final push by IKN to try to save the bidding process. When news of IKN's renewed bid bond became public, IPC and the Chaudhary Group screamed in protest to the press—and threatened to withdraw their bid—claiming (probably correctly) that the lapsed bond should have disqualified IKN from the bid. Then a December 10 *Kathmandu Post* article reported that the ministers of water resources and finance had met and agreed to cancel the privatization bid and call for retendering. When Pradhan and Jantzen went to the Privatization Cell that day to ask what had happened, the people there could only say that none of them had made any such decision!

Hoping that there might still be some small spark that could be fanned to bring the privatization process back to life, Hoftun got on a plane for Nepal. Maybe a visit from the grand old man of hydropower in Nepal might sway the powers. Along with Pradhan, Hoftun managed to go straight to the top with a meeting with Prime Minister Bhattarai. For the umpteenth time the two laid out the case for a speedy resolution to the privatization stalemate in order that the Melamchi project could get underway. At the end of their presentation the prime minister told Hoftun, in English, "I will support you," before turning to the others in the room and saying in Nepali, "But the problem is that Mahesh Acharya [the finance minister] will not listen to me!"

This confirmed earlier reports that both the ministers of water resources and housing and physical planning were in favor of the IKN bid, not least because of BPC's historic links to Norway and Norad's desire to include BPC in the Melamchi project. They both favored the project *with* the powerhouse component. But the finance minister was opposed. He argued that the current bids for BPC were far too low, barely half the USD 20 million that the valuator had estimated as BPC's worth. Retendering, he argued, would bring in higher bids and more money for the government. As for the Norwegians, he noted that neither the embassy nor Norad had ever *explicitly* tied their Melamchi support to IKN's bid for BPC. Pradhan's source reported the minister as having said that, unless the Norwegians provided something in writing expressing their support for IKN, he was not going approve their bid. Plus, the minister feared that selling BPC for the current bid prices would only invite heavy criticism from the opposition. The Nepali Congress did not want to be seen as handing over valuable government assets to foreigners for bargain prices.

In desperation Hoftun and Pradhan paid yet another visit to Norwegian embassy officials, the kind that might actually put "something in writing" (regarding Norway's wishes that BPC go to IKN) that would sway the finance minister. They got a frigid reception. The embassy's Norad representative, who for months had been increasingly skeptical of the Melamchi hydropower component, now flatly rejected it. According to Hoftun's notes on the meeting, the Norad official said that "the future ownership of BPC had nothing to do with Melamchi. The tunnel can be constructed without a power plant." Furthermore, she added, "There is no requirement from Norad's side that Nepali companies should do the work," thereby snubbing one of IKN's main contentions (and apparently contradicting the language of the original Norad Melamchi grant). As if to rub it in, she then accused IKN of ruining their chances on the first bid "first by bidding too low, and then by not renewing their bid bond on time." To her, IKN "had proved not to be competent in international business, and now there was nothing more to do about this." When Hoftun asked if she felt no responsibility to nurture Norad's thirty years of investment in Nepal's independent hydel sector, Hoftun paraphrased her answer to: "she could not care less." Later Pradhan told Hoftun that, in earlier meetings with this Norad official, she had dismissed the matter by noting that "Mr Hoftun is sentimental about BPC." Certainly she was not. Hoftun could only conclude that Norad no longer saw "the industrial-engineering sector of the economy ... as part of development." Norad, along with Statkraft and other Norwegian entities, seemed to have jumped onto ADB's bandwagon along with its big-capital, big-finance, essentially predatory "development" practices.

Hoftun left Nepal in late December without having accomplished much to break the stalemate. The prime minister and other top officials favored the IKN bid but the finance minister was staunchly opposed, favoring retendering. With the support of the Privatization Cell and its British advisor Clarke, Pradhan and Hoftun played their last card which was to have the government "call the bidders for negotiations as per normal procedures and then award the tender to the bidder which had offered the overall best bid." According to Clarke, retendering would be a "disaster" that—given the fiasco of the first tender—might result in lower bids, or even none at all. Better to bring things to a close with the best possible outcome for all concerned.

Hoftun let it be known that it was very unlikely that IKN's Norwegian investors would be willing to bid a third time, given the government's poor handling of the process. But even more to the point, with the Melamchi power plant looking less and less likely—and given that BPC's at one time seemingly guaranteed role in that hydel project was an important positive factor in the eyes of potential bidders—Hoftun knew that BPC without the Melamchi contract would be much less attractive to foreign investors. For IKN to buy BPC, BPC needed Melamchi. Now IKN looked to be on the verge of losing both.

Even worse, from the government's perspective, if Norad was *not* making its financial contributions to the Melamchi project conditional on either BPC's participation or a hydroelectric component, then the whole matter of BPC's privatization became much less of a priority. Why not shelve BPC privatization altogether and prioritize getting Melamchi underway as a straightforward drinking water project?

In spite of widespread rumors that the privatization process was dead, on January 27, 2000, Balaram Pradhan got an early morning phone call from the minister of commerce informing him that BPC privatization was finally on the cabinet's agenda for that day, and that he needed clarification. The minister's main question was why the bids were so low, much lower than the company's valuation. Again Pradhan had to explain the most basic points, such as the fact that the government was selling only 75 percent of the company shares that it held, therefore only 75 percent of the business was being purchased. Again he stressed the need for or lack of a PPA, without which there was no way of gauging BPC's profitability looking forward. More than anything, Pradhan pleaded that the simplest, most logical thing to do was to simply bring the two bidders to the table, let them negotiate the best price and conditions, and be done with it. But already later that day Kathmandu newspapers proclaimed "HMG Preparing for Retendering of BPC" with the justification that the bids were too low for a profitable company. The second round of tendering was dead and BPC was still firmly in government hands.

Still in play, though barely, was the possibility of a Melamchi project that still included a hydropower component. Officially the Norwegians stuck by the feasibility study they had commissioned earlier from Norplan and BPC Hydroconsult—the design with a higher elevation tunnel, a 250-meter drop, and a power-generating component. But ADB, which had opposed the power plant from the beginning, hired the Japanese engineering consultancy Nippon Koei to carry out another detailed feasibility study for the project without the power plant. In a marathon two-day meeting in Kathmandu in early February 2000, technical experts from Norplan and Nippon Koei went head-to-head, with a surprising amount of disagreement on even the most basic points. For example, Norwegian test drilling had found rock conditions to be "very bad" whereas the Japanese had found conditions to be "very good." And while the Japanese wanted a relatively small free flow tunnel of only 6–8 square meters (at the face), the Norwegian design called for a much larger (12 square meters or greater) tunnel running under pressure. Both sides claimed their design would be more efficient to build and easier to maintain.

A week later Balaram Pradhan reported excellent news to the IKN consortium that, at the minister of water resources' insistence, the government had chosen the Norplan–Hydroconsult design. Even better, the same minister was actively taking steps to lease new distribution areas to the future owners of BPC. There was even serious talk of starting PPA talks with NEA—even though NEA employees and their union remained dead set against any move that would increase the likelihood of BPC's privatization.

But the IKN team barely had time to savor their good fortune before their hopes came crashing down again. On February 20, 2000, yet another cabinet reshuffle forced numerous ministers to resign, including the very pro-IKN, pro-BPC water resources minister who had appeared so recently to have broken the logjam. Perhaps sensing a now softer target, in early March ADB sent a strong letter to the Nepal government stating unequivocally that the power plant was, according to their experts, fundamentally incompatible with a water delivery system and therefore must go. In fact, ultimately ADB placed the blame for rejecting the hydel project on the Nepal government, arguing that in Nepal any project that fell within the jurisdictions of more than one ministry would be a disaster: Nepal was incapable and unworthy of a complex project. For Hoftun and Pradhan ADB's intrusion was an ominous déjà vu replay of the Khimti project's hijacking and subsequent cost escalation. Desperately Hoftun complained that "getting ADB involved is just calling for confusion." But Pradhan warned that "HMG is watching the attitude of the stake holders," that is, watching to see which of the donor agencies was most powerful or at least most persistent.

ADB announced a donors' meeting in Manila in late March where it was assumed they would pressure Nepal government officials to cancel the power plant. Norad and other Scandinavian aid groups refused to attend, arguing that there had been enough meetings already. At this point Nepali officials bowed to ADB and its money, allowing them to commandeer the Melamchi project and take the lead on its construction. BPC and its daughter companies had now been frozen out of the project that they had taken the lead in developing for decades.[21]

ADB went ahead with international bidding for the tunneling job as promised. As expected, a Chinese firm won the bid only to—also as expected—fail miserably due to underestimating the challenge and having bid too low to carry out the project. Construction incompetence combined with the chaos of the Maoist People's Rebellion caused repeated delays which, in turn, inspired many of the project's initial backers to simply pull out. Eventually money from the Organization of Petroleum Exporting Countries (OPEC), the Japan International Cooperation Agency (JICA), and others allowed tunneling to recommence with an Italian construction firm—which also eventually pulled out of the project. Tragically, just as the Melamchi tunnel was finally nearing completion, in 2021 a massive landslide buried the headworks leaving the fate of the entire project in limbo.

We will never know whether an IKN-controlled BPC could have completed the project in just five years as promised (as opposed to the more than twenty it has now taken). But it is almost certain that they would have built the tunnel and power plant largely with Nepali skilled labor and NHE's Nepal-built components for a fraction of the cost that resulted from ADB's hostile takeover of the project. Experienced and knowledgeable about

---

[21] See Pokhrel (2017: 180–183) for a brief survey of the Melamchi project's history.

everything from rock conditions to political conditions to social and labor conditions, BPC would have likely weathered the challenges of construction more efficiently than the foreign contractors who have floundered and wasted huge amounts of money. Even more importantly, BPC and its corporate progeny would have emerged more empowered, more experienced, and in a position to lead the hydropower development sector in Nepal and even beyond.

Instead, hundreds of millions of dollars went to foreign firms while skilled Nepali labor continued to flow abroad and BPC continued to languish under government mismanagement for still more years as the privatization process dragged on and on.

7

PRIVATIZATION

The Long Haul

You may say one thing more: Mr. Hoftun is not going to give up.

—Odd Hoftun to Balaram Pradhan, January 28, 2000

The Butwal Power Company (BPC) privatization process went through four rounds of bidding. The Nepal government canceled the first two rounds officially on the grounds that the bids were too low. But, as we have seen, there were many other complicating factors including negotiations surrounding the Melamchi project and active obstructionism on the part of the Nepal Electricity Authority (NEA) and the BPC administration itself, both of whom feared that the company's privatization would threaten their authority and even their jobs. Adding to the climate of dysfunction was the general state of chaos that had enveloped Nepal during the People's War when the government was in constant flux and *bandh*s (public strikes) and bombings became almost routine threats to daily existence.

The third and fourth rounds of bidding turned out to be very different from the first two. Whereas the earlier rounds had seen significant international interest in purchasing BPC, by the last two rounds virtually no foreign bidders remained. A company that had earlier been portrayed as a promising international investment opportunity had, from a Western perspective, turned into a third-world albatross. Ironically, as foreign interest dwindled, Nepali investors stepped into the gap to finance almost the entire bid.

Looking back, perhaps the most astonishing thing about the entire process—beyond the ongoing obstructionism and chaos in which it unfolded—was the sheer determination of the bidders. Both the Interkraft consortium and the Chaudhary Group held out till the bitter end in what looked like a war of attrition, or an ultramarathon that few mortals could have succeeded in finishing. By the time it was over Odd Hoftun was emotionally and physically exhausted and Balaram Pradhan had, literally, worked himself almost to death.

In the weeks following the second bid's cancellation, in January 2000 Hoftun and Pradhan reflected on where they had been and where they were going. Both sensed that even the limited Norwegian public-sector commercial interest that BPC had attracted in the first rounds would not continue into the third. Although it added a great deal of stress to the prospects of financing the remaining bids, in a way this stripping away of foreign commercial involvement was liberating for Hoftun and Pradhan. Neither man had ever understood profit to be BPC's reason for being. Foreign commercial ownership would have severely tested Hoftun's commitment to reinvestment of profits into corporate growth and capacity building in Nepal's hydropower sector. But with foreign commercial investment now out of the picture, Hoftun, Pradhan, and others could focus on what had truly motivated them from the beginning: shaping the company's future direction. Hoftun knew that he would likely lose the kind of control that majority ownership would have brought. But he felt that he could influence the company by influencing its management after years of government neglect. "My concept of the future BPC is so much dependent on putting in a new management team who will manage the company in a positive manner." More than anything Hoftun wanted "to make sure that good dependable new owners take over BPC and manage the company in a manner which is truly beneficial for Nepal." The day after the government officially announced retendering for a third bid (January 27, 2001), Hoftun wrote to Pradhan that anything may happen, "but we cannot give up."

THE THIRD BID

One thing that did not change in the final two bidding rounds was the process's nerve-wracking tempo: long periods of government inaction and postponement, followed by short bursts of intense activity as bidders scrambled to meet arbitrary deadlines or swerve around the latest obstacle thrown in their path in hopes of wrecking the process altogether.

The first wait was simply for the new bid documents. By March 2000 came news that the Privatization Cell (which had managed to retain its British legal advisor Douglas Clarke) had almost finished preparing the bid documents but was waiting for official action on the same two old sticking points: a new power purchase agreement (PPA) and some decision regarding a new distribution area for a privatized BPC. Over the following months came occasional promising news from the ministry of water resources (MoWR) on these topics but each time NEA moved to block or delay any progress. For example, even after a new "ten district" plan had been approved at the ministry's top levels, the NEA board chairman demanded that, before any further action, the existing (government-owned) BPC staff needed to prepare a detailed report laying out exactly how the retail transfer would be carried out, what would happen to NEA staffers, and so on (as though the current government-owned BPC knew what a privatized BPC would do). When informed of this in a meeting with officials, even the normally unflappable Balaram Pradhan lost his cool and, "speaking loudly and probably furiously" (as he reported to Hoftun), berated everyone in the room for sabotaging Nepal's domestic power industry and dooming the nation to forever go begging for money and jobs to the likes of the World Bank and the Asian Development Bank (ADB). He reminded them that Nepal's current five-year plan explicitly called for the privatization of electrical distribution and accused everyone of stalling. With the bid documents seemingly endlessly delayed, Hoftun complained that it was almost impossible to attract new foreigner investors—people who were not going to commit to *anything* unless and until they had a very clear sense of what was being sold.

When in August 2000 the Privatization Cell finally released the bid documents, six months after the bid had been announced, it was without *any decisions* having been made regarding a new PPA or distribution area. If anything, the documents represented steps in the wrong direction, not just stasis. For example, not only had the PPA not been renegotiated or even simply extended, but the "take or pay" clause of the previous PPA had been revoked. In other words, NEA was no longer even obliged to purchase power produced by BPC plants. Hoftun moaned,

> It is rather incredible, exactly the same as last time. Nine months from now it expires.... And still HMGN wants a higher price than before! It is now clear that NEA will buy power from BPC only when they need it, and take nothing when they have enough power available themselves. No power producer can live with that kind of PPA.

Hoftun could only conclude that NEA's new aim was "to kill BPC as a competitor." "This really spoils whatever credibility HMGN still had with foreign investors," Hoftun wrote to Pradhan, "at least here in my circles in Norway."

Perhaps the most unsettling news contained in the third bid documents was a specific reference to a contentious income tax assessment that had been hanging over BPC for years. At question were taxes on profits from BPC's Andhi Khola power plant. Because Andhi Khola was commissioned shortly before the 1992 Electricity Act went into effect, there was debate between government agencies regarding how Andhi Khola should be handled. According to the Act, new private power plants enjoyed a fifteen-year tax holiday. MoWR held that this tax holiday should apply retroactively to Andhi Khola and continue through its first fifteen years. But tax authorities argued that current law *should not* apply to Andhi Khola, which should pay its income tax for the period since it began operation. The new bid documents included the claim that tax authorities had assessed income tax (roughly NPR 150 million) on BPC to retroactively cover the previous eight years of production at Andhi Khola. On top of this came "non-refundable penalties amounting to NPR 12 million and 50 percent deposit for tax assessed so far." The newly assessed total income tax liability was equal to about one quarter of BPC's entire book value!

These charges had already been brought before a tax tribunal but no one expected a resolution anytime soon. If anything, the case was likely to go to the Supreme Court and drag on for years. Hoftun complained that the whole matter had been thrown into the gears in an effort to sabotage the privatization process. "It seems very unfair," he wrote, "to leave this issue for the bidder to sort out when it could have been settled internally between the two concerned departments of HMGN long ago." Instead, it has been thrown into the proceedings "like a time bomb for the winning bidder to defuse. Certainly, it would have a positive effect on the bid price if this issue would be settled beforehand." Hoftun called for a clause in the privatization agreement whereby the government would promise to compensate the bidder if the matter was decided against BPC. Until that time, the tax issue would sit like a loudly ticking bomb in the middle of the BPC privatization package.

If the tax issue was a threat designed to scare away bidders, there was no shortage of other frightening developments—including *real* bombs—to discourage foreign investors. Nevertheless, Hoftun's next job was to try to sell BPC, under its current conditions. If anything, those conditions were only deteriorating—literally. In May 2000 Hoftun visited the Jhimruk power plant and was appalled at what he found. The project's severe sediment and erosion problem was nothing new but Hoftun learned that, under government management, the Jhimruk staff had neglected routine maintenance to the point where the power plant would need to shut down because of damaged equipment. Worse still, not only were there no spare turbine parts on hand, the government had not even budgeted for them—leaving the power plant in a condition where it would be forced to shut down with no spare parts and no money to buy them. This budgeting failure came in spite of the fact that the government-controlled BPC board had issued—basically to itself—a 15 percent stock dividend the previous year.

By August 2000 the plant had indeed been shut down for several months. Pradhan complained that "the company is running down every day" and "may go into loss next year." When Pradhan informed high officials in MoWR of Jhimruk's condition they expressed great surprise and refused to believe his reports. Pradhan could only conclude that the government itself was "completely misinformed ... about the BPC situation." Of course the Jhimruk power plant's physical problems were only compounded by the project's intense social problems and local unpopularity (described in Chapter 4). At the height of the People's War community relations at Jhimruk had gone from bad to worse. The business plan that IKN submitted for the third bid no longer even considered Jhimruk to be an attraction: "It has to be seen as a liability rather than an asset."

Between having no PPA, no distribution agreement, huge threatened tax assessments, infrastructure deterioration, civil war, political chaos, and other problems, BPC was hardly an appealing investment opportunity.[1] "One cannot go out and talk very enthusiastically about investments in Nepal when the news from that end is so bad," Hoftun complained. After a year of perpetually shifting governments, debilitating public strikes, bloody violence in the districts, and bombings in the capital, even Balaram Pradhan confessed that "we are wondering about the future of the nation."

Even the language used in the government's third bid documents reflects a negative shift in perspective regarding BPC's prospects. Rather than a company with a proven track record promising relatively safe investment and short-term profit potential, in the third bid BPC had become an opportunity for "strategic investment"—a company with (possible) long-term potential. As Hoftun noted, strategic investment called for "risk capital," something that few are interested in allocating in large amounts and is anything but reassuring for the average profit-driven investor. In other words, labeling BPC a "strategic investment" was an admission that not only will profits be limited and late (if ever) to arrive but that the whole prospect is riskier than previously portrayed.

With this shift, "Norwegian investor attitude has almost totally changed," reported Hoftun. Initially the first bid had attracted significant direct foreign investment. But even after most of the major commercial Interkraft investors had backed out, two private Norwegian utilities (SKK and Agder Energi) had stuck with IKN through the first and second bids. Now, with the third bid, they too dropped away. SKK was the first to go, its representative having told Hoftun "quite bluntly that he had had enough of Nepal." For a while it looked as though Norfund would match a small investment from Agder

---

[1] As if to make matters even worse, a September 21, 2000, *Kathmandu Post* article reported that moves were afoot among politicians to "revise" the deeply unpopular Khimti PPA with Himal Power Limited (HPL) and its main shareholder Statkraft. While revisions might well have been justified, this kind of news was enough to unnerve even the boldest investors because it suggested that even formal legal commitments between private investors and the Nepal government were potentially subject to change according to the whims of politicians.

Energi but that too fell through when it became clear that Agder Energi's precondition of 51 percent foreign ownership was not going to be met. For a few months Hoftun courted the Australian developer Pacific Hydro, who showed enough interest in BPC to commission a Price Waterhouse Coopers assessment of the company. Unfortunately, the report threw up so many red flags that Pacific Hydro quickly backed out. Desperate to maintain at least some foreign commercial involvement in the Interkraft Nepal (IKN) bid, Hoftun convinced Agder Energi to stay with the bid, though now only as a corporate *contractor* providing technical services to BPC on a commercial basis. The foreign expertise and technology transfer that IKN had once assumed would come from willing investors had now become something available only through purchase.

With regard to foreign investment, by late 2000 Hoftun wrote that "we have had to conclude that we may have to do it alone with our Nepali partners." But if IKN was named for the Interkraft International group of Norwegian utilities that were IKN's original investors, who was the "we" that Hoftun referred to now that foreign commercial investment was a thing of the past? In fact the name "Interkraft Nepal" lived on even after IKN's composition, ownership, and role had completely changed. By September 2000 Hoftun could report that "I believe we do now have in place the foundation for a credible bid, but without committed private investors. This new foundation still bears the name Interkraft Nepal AS (IKN)." But now a Norwegian trust—known as "Stiftelsen Hjelp til Selvhjelp for Nepal," or just "Stiftelsen"[2]—owned 100 percent of IKN's shares. Once it took over the BPC bid, Stiftelsen reconstituted itself with a new board made up of people with previous Nepal experience, with Hoftun as its executive manager.

The new Stiftelsen-controlled IKN was no longer simply a means to carry out a bid for BPC. Because it no longer represented "committed private investors" from Norway, the new IKN—with Hoftun now firmly in control—now understood itself to be "a more permanent organization which

---

[2] Hoftun's mission organization, the Norwegian Himal Asia Mission, had set up Stiftelsen some years earlier to oversee and facilitate various economic development projects in Nepal. Technically, Stiftelsen purchased IKN from Interkraft International (the Norwegian utility consortium) for a nominal price.

would offer management and engineering services to BPC in case IKN wins the tender." In addition to its own in-house engineering skills, the new IKN would now act as an intermediary between a privatized BPC and Agder Energi, the Norwegian firm that had agreed to provide development services to BPC on a paid contractual basis. Now led by people with long Nepal experience, Hoftun was confident that the new IKN would be able to act in a more efficient and decisive way than when it had represented more timid, slow-moving corporate interests.

But what remained anomalous about the new IKN was that even though it no longer represented the lead investors—a role now taken over by the Nepali side of the consortium—the Norwegian owned and controlled IKN still retained the role of "consortium leader" and officially designated "strategic bidder." Hoftun and the IKN board were the official leaders of the BPC bid effort even though the vast majority of the actual investment funds were coming from the Nepal side. Of course this was logical in many ways, given that Hoftun had deep Nepal experience and had initiated the entire bidding effort. But this dynamic also points to a stark internal leadership contradiction between the Norwegian "consortium leaders" and the Nepali investment "leaders"—the people who were actually putting down the money to buy BPC. In what was perhaps the first hint of things to come, already in the weeks following the collapse of the second bid, one of the Nepali investors had inquired about the possibility of their taking over the bidding procedures. But others on the Nepal side were less interested and Hoftun (whose main motivation was to have a hand in guiding a privatized BPC) was opposed. As we will see, tensions steadily built along this fault line until it acrimoniously split apart years in the future.

Ironically, as foreign investor interest in BPC dropped away, interest among Nepali investors surged. This is partly because the third-round bidding documents no longer required investors to have prior hydropower development and management experience. But just as importantly, Nepali commercial interests were beginning to understand the nature and potential of hydropower investment. Khimti, the first privately developed power plant in Nepal, was up and running and had paved the way for others to follow. And the fact that the Marwari Chaudhary Group was involved only added fuel to the fire by stoking nationalist sentiments. Already in February 2000

Gyanendra Lal Pradhan, the de facto leader of the Shangri-La Energy Limited (SEL) Nepali investors group, complained that "there are too many investors interested!" The original seven SEL investors from the earlier bids considered joining forces with other Nepali business houses but, in the end, outside interest in the bid only encouraged them to keep the matter in their own hands.

Whereas in the first two bidding rounds IKN had functioned as a bidding consortium bringing the majority Norwegian investors together with the minority Nepali investors, in the third bid IKN represented only the Norwegians while SEL represented the now-majority Nepali investors. For legal purposes these two entities formed a joint-venture company called Interkraft Holding Ltd. that would be the official bidding consortium, even though IKN retained the role of "consortium leader" and official bidder.[3] Representing each side's contributions to the bid total, IKN held 14.3 percent of Interkraft Holding Ltd.'s shares, and SEL 85.7 percent. Reflecting this unequal division, the joint-venture agreement stipulated that Interkraft Holding Ltd.'s seven-person board would have five members appointed from SEL and two from IKN. To (hopefully!) reduce confusion, in what follows I will continue to refer to IKN as the joint bidding party—as the participants themselves did—while when necessary distinguishing IKN Norway (the Norwegian minority bidders) from the Nepali SEL majority investors.

In its basic structure the third bid was similar to the first two: a "two envelope" system requiring separate technical and financial bids. As in the earlier rounds, evaluators scored the technical bids out of a possible 100 points based on the following conditions:

- Bidder's capability: 40 points (proven technical capability, 20; proven financial capability, 10; proven development capability, 10)
- Company's reputation: 15 points
- Future plan and proposals for BPC: 30 points
- Contributions to Nepal's financial development: 15 points

---

[3] IKN retained the official leadership role in order to fulfill the technical experience and competence required in the bid documents.

Without its big foreign backers, IKN's bidding position was much weaker than in the first two rounds, but so also was the competition.

For the third-round technical bid Hoftun submitted exactly the same proposal as in the first two but with "fresh observations" inserted in italics to designate changes from the earlier version. For example, where the earlier bid had discussed BPC's planned role in the Melamchi project, here Hoftun adds:

> The Melamchi hydropower project was killed by one of the International Banks. In Bidder's opinion this was a regrettable loss to the country.... The fate of Melamchi is one of many examples illustrating how speculative the launching of specific plans for new hydro projects is under the prevailing circumstances. Too many grand ideas are floating around. It may be better to present more realistic plans for the future.

Whether this rejection of "grand ideas" is a critique of IKN's own earlier grand Melamchi plans (diverting multiple rivers, multiple power plants, and so on) or of Independent Power Corporation's (IPC's) promises to invest huge amounts of money in Nepal is not clear. But this back-to-the-basics, embrace of "realism" pitch seems slightly duplicitous given that IKN's now rather enfeebled financial condition would have made any other approach impossible. Nevertheless the whole bid has a much toned-down quality. "This Bidder is not going to boast about big plans for mega-projects," says Hoftun. In the third bid promises for growth through direct investment are gone. Now the plan is to grow through "increased efficiency" and "good and honest work." With language like this we are definitely back in the old Hoftun-inspired socio-moral development universe.

Through the fall of 2000 and into the early winter of 2001 the government repeatedly postponed the date of bid submission. In the midst of ongoing political turmoil, Maoist attacks, *bandh*s, and repeated efforts to scuttle the plan, the whole privatization process seemed at times to be dying a slow death. At the Privatization Cell Pradhan could not even find anyone to talk to aside from Clarke, the lonely and despondent British advisor. In the meantime IKN watched as its few remaining European backers dropped out, leaving it with only about USD 350,000 to contribute to the larger effort. "There is no one to be blamed except the situation in Nepal," Pradhan wrote consolingly. Afraid that "those will win who want to see BPC disappear, or be absorbed into NEA," Hoftun asked: "Isn't this a cause worth fighting for, to provide competition for NEA even though it is steadily becoming less interesting for investors to take over BPC?" Of course for Hoftun the answer was yes, and always would be.

By late January 2001 the situation had become so unpredictable that Binod Chaudhary, the leader of IKN's earlier nemesis the Chaudhary Group, approached Balaram Pradhan with an offer to merge the IKN and IPC bids such that each would hold half of a 51 percent controlling interest. Hoftun was tempted by the offer reasoning that if IPC put in some money, some of his Norwegian investors might rethink their decisions to pull out. But some of the Nepali investors surmised that Chaudhary's new attitude might be the result of IPC having backed out of the bid, leaving him with no partner. In the end the Nepali investors decided that it was too risky a proposition. Still, the fact that these arch enemies were willing to dance together if only for a moment suggests how dire the financial conditions had become.

When the date for bid submission finally arrived on February 14, 2001, the occasion turned out to be quite dramatic. The afternoon deadline came and went, leaving the IKN bidders jubilant that they had won the bid by default. But then, fifteen minutes late, a Chaudhary Group representative arrived bearing bid documents. In spite of their rule violation, ministry officials accepted the Chaudhary bid while promising that "necessary action will be taken at the appropriate level" to penalize them. This promise was none too reassuring to many on the IKN side who had long been convinced that finance ministry officials, from bottom to top, were thoroughly bribed by the Chaudhary Group. Several in the IKN group demanded that Balaram Pradhan (who was acting as IKN's representative as lead bidder) protest by refusing to sign *anything*, reasoning that signatures would only validate the shady proceedings. Switching to Newari the IKN group had a heated discussion with the protest advocates arguing that if Pradhan signs the papers, and if the ministry refuses to penalize the Chaudhary Group, and if the other side's financial bid is higher, IKN will have lost everything. But Pradhan argued that by refusing to sign, IKN itself risked being accused of not following prescribed protocol and therefore being disqualified.

Pradhan went ahead with signing and was glad he did. Part of the requirement was for Pradhan to initial the back of each and every page of the Chaudhary Group's bid documents. In so doing Pradhan had the chance to quickly surveil the other side's bid and found that, indeed, the Chaudhary Group's British partner, IPC, *had* backed out. Among the documents were brief letters of interest from Australian and Scottish investors but none of the required joint-venture documentation. Furthermore, the Chaudhary Group's application had neither the required cover letter nor business plan. Having seen their opponent's bid, Pradhan could report that it was very, very weak.

IKN's total financial bid in the third round was NPR 730,000,001 (c. USD 10 million),[4] with the Norwegians providing only USD 350,000 (NPR 25.7 million or 14.3 percent) and the Shangri-La Energy Ltd. Nepali investors the remaining NPR 704 million (85.7 percent). This came to NPR 116 per share. The Norwegian side's contribution may have been much smaller, but at least it was cash in hand. By contrast the Nepali side offered nothing but promises that bank loans would be forthcoming if needed. Hoftun had been in Nepal during the bid submission period but when he returned to Norway in late February he reported unsettling news. His fellow Norwegian IKN board members were

> unhappy about the deal we made. They feel that I had gone too far in accommodating to the financing arrangements our Nepali partners made, or rather have not made. People in Norway cannot understand the business practices of Nepalese trading houses. They think that it is an irresponsible act to hand in a bid on such a basis. They want to see things on paper, and not just hear my verbal report quoting them saying that it will be easy to raise the money if and when the IKN bid is selected. This difference in thinking is the same as what I experienced also when SKK dropped out. The confidence is lacking.

Having seen so many foreigners lose heart already, Hoftun feared that even the little foreign investment he had been able to scrape together for the third bid might disappear.

---

[4] Unlike the earlier two, the third bid was to be made in Nepali rupee terms.

Once bids were in it was time, once again, to wait. The ministry's promise to open financial bids in two weeks proved to be meaningless with no end in sight. The following months saw war atrocities like the Rukum Massacre, right-wing calls for a royal coup, and then, in early June, the horrific Palace Massacre. In shock, Pradhan wrote to Hoftun, "It looks as if we have no Government functioning in the country."

Worse still for the privatization process, (the now infamous[5]) Lokman Singh Karki had been appointed secretary of MoWR. As water resources secretary, Karki was also head of the government-controlled BPC board and was making it increasingly clear that, as Pradhan reported, he "would never support privatizing BPC"—a fact that the Privatization Cell's Douglas Clarke agreed was "obvious." Pradhan accused Karki of sitting on paperwork for months and "causing unnecessary delays." What's more, Karki was actively feeding disinformation to his superiors up the chain of command, claiming that BPC was in strong financial condition, was a big profit maker for the government and, therefore, should not be privatized at all. Karki told his superiors that IKN's bid price was not even enough to pay for BPC's shares in Himal Power Limited (HPL) (the owner of the Khimti plant), let alone for the whole company. It was all Pradhan could do just to convince high officials—the ones who had to make actual decisions—of BPC's actual state of affairs and the necessity to push privatization forward.[6] Sensing that he was fighting an uphill battle, Pradhan confessed to Hoftun that "I have very little hope that BPC will be given to IKN."

In early August the finance ministry finally announced that the Chaudhary Group had been disqualified because of an inadequate technical bid score,[7] and that the financial bids would be opened by mid-month. Incensed at

---

[5] See Kshetry (2018).

[6] Ironically it is at this stage (mid-July 2001) that NEA decided to make one small concession to the BPC privatization process. On July 13 NEA announced that the existing PPA for BPC's two plants would be extended for two years. Given that this move came *after* the third-round bids were in, and that a two-year extension hardly made any difference in terms of long-range financial planning, it's not clear why NEA made this move.

[7] Without the official involvement of IPC, the Chaudhary Group did not have the necessary experience needed to qualify for the bid (Bhattarai 2001).

having been thrown out of the bidding, Binod Chaudhary launched an intense media disinformation campaign. Among other things, he claimed still to represent IPC even though IPC had dropped out of the bid.[8] Pradhan complained that the *Kathmandu Post* and *Kantipur* both published articles extremely favorable to the Chaudhary side while refusing to publish factual material provided by IKN. Pradhan accused the papers of bowing to pressure for fear of losing major advertising revenue from the Chaudhary Group's many local businesses.

After several postponements, the ministry of finance set the financial bid opening for Friday noon, August 17, 2001. According to the ministry's directives, a Chaudhary Group representative was to be present to take the bid documents and leave the room before the opening of the qualified bidder's (IKN's) documents. But the press reported that the Chaudhary Group was threatening to remain in the room and publicly open their financial bid if theirs was higher than IKN's. In front of assembled high officials, the finance minister announced IKN's bid to be NPR 730 million. Then, as promised, a Chaudhary Group representative publicly announced that their bid had been NPR 820 million, though he offered no proof (Bhattarai 2001). The finance minister then proclaimed that, having been technically disqualified, the Chaudhary Group bid was irrelevant anyway. But rather than accepting the IKN bid, the minister announced that they would evaluate the bid and make a decision "within a week."

Three weeks later there was no decision but Pradhan reported rumors that the water resources minister and secretary had decided that IKN's bid was too low and that they would call for rebidding. Cynically Pradhan concluded that the two had not received adequate "personal benefits" though officially the ministry claimed that it had found a "serious mistake" in the IKN bid document. When pressed this turned out to be a typo in the opening paragraph where a 7 had been written instead of a 6. The Privatization Cell countered that this was an easily remedied error and certainly not grounds for the bid's rejection. By September 10 Pradhan reported that completely

---

[8] It later turned out that IPC was to be a paid consultant for the Chaudhary Group, not a stakeholder. This was not unlike the relationship between IKN and Agder Energi, though Chaudhary did not portray it that way.

contradictory reports were circulating. Some ministry insiders said that the bid had been cancelled and rebidding would be called. But Douglas Clarke of the Privatization Cell told Pradhan that a proposal to grant the bid to IKN had gone up to the cabinet level. "Everyone is totally confused now," said Pradhan.

The next day the press reported that the bid had been canceled for being too low and for minor anomalies. The recommendation that had gone to the cabinet was for retender, and they agreed. Bid three was dead. As news of the bid cancellation sank in, Gyanendra Lal Pradhan told his nephew Balaram that he'd had enough of the whole BPC affair. "I just told him that Hoftun may not get tired so soon. We can think and plan for the future."

THE FOURTH BID

The official call for a fourth round of bidding came on October 11, 2001, with instructions that "fresh bids" be deposited on November 20. Given that all the previous bids were rejected as too low, this time there was at least some recognition of bidders' concerns in hopes of raising the Butwal Power Company's (BPC's) value in the eyes of investors. Although neither the tax issue nor the distribution area matter were resolved, at least with the fourth bid documents came welcome news of a ten-year extension of the current power purchase agreement (PPA).[9] But there were also discouraging developments. BPC's current PPA required that it run its Jhimruk plant at full capacity

---

[9] One of the sticking points regarding a new PPA for BPC's power plants was the concern that BPC was making too much money selling electricity. According to a hydropower expert I spoke with, the Nepal Electricity Authority (NEA) normally based its purchase rates on the assumption that private developers would have invested around USD 2,000 for every installed kilowatt of power production capacity. (That is, they assumed a 1 MW plant would cost about USD 2 million, 5 MW 10 million, and so on.) Purchase rates had to be set allowing a fair profit to investment ratio. BPC's "problem" was that, by using labor intensive methods, used equipment, and other cost-cutting measures, BPC plants ended up costing only about USD 1,000 per installed kilowatt. Therefore, NEA was very reluctant to set its power purchase rates for BPC at the same level as it did for other commercial projects arguing that, if they did so, BPC

during the monsoon but by the early fall word came of exceptionally heavy turbine damage due to sediment erosion. It was becoming clear that the most economical route for Jhimruk would be to greatly reduce or even entirely shut down operations during the monsoon with the loss in power sales being less than the cost of damage sustained. All of this meant that Jhimruk's profitability was dropping and with it, the value of BPC as a whole. This was *in addition to* the steady decline in profitability that BPC had experienced since it came under government control (Bhattarai 2001).

But 2001 also brought some encouraging news for BPC and possible investors. At the end of fiscal year 2001 Himal Power Limited (HPL)—the owner/operator of the Khimti power plant—issued its first financial earnings report, and the news was good (for stockholders). Khimti had generated a net profit of USD 7.7 million (paid *in USD* thanks to the extremely favorable terms of Khimti's PPA). Even though BPC had been reduced to holding only 15 percent of HPL's shares, this news meant that, in the future, BPC could expect to earn *at least* USD 1 million per year in profit from Khimti alone. This news made many investors, both foreign and Nepali, take another look at BPC. As one Nepali observer noted of potential investors, once HPL "distributed a handsome dividend, then their eyes got wide!"

Yet again Hoftun set out to raise investment capital in Norway but the new bid date gave him only a month to work with. This time he got encouraging signals from Agder Energi officials but they could only come to a final decision at their next board meeting, which was one day *after* the November 20 bid deadline. Pradhan and Hoftun asked the Privatization Cell for a two-week delay—arguing that it might allow them to raise their bid price—but were "flatly refused." Once again the Nepali investors agreed to shoulder the majority of the financing but they also made it clear that they would still be happy to take a backseat to Norwegian majority investors. A week before the November deadline, Shangri-La Energy Limited (SEL) Board Chairman Padma Jyoti wrote to Hoftun encouraging him to raise more Norwegian investment capital. "Our view is that our strength to manage BPC successfully will increase because of this," Jyoti wrote. "However, we

---

would be making an unfair profit. But BPC felt it should be awarded the same rates as other private power producers, rather than being punished for its own efficiency.

propose that Norway should take not 51% but 65% of shares, Nepalese will take 35%." The SEL investors were eager to make the bid successful but were also eager to limit their risk exposure.

Ultimately the finance ministry extended the bid deadline to December 5, 2001, to accommodate an investor group that had only purchased the bid documents at the last minute. Also discouraging for the Interkraft Nepal (IKN) side was news that the British Independent Power Corporation (IPC) had renewed its joint venture agreement with the Chaudhary Group. In the new joint venture company (Independent Power Corporation Nepal, or IPCN) the Chaudhary Group controlled 80 percent of the shares and the British firm only 20 percent.

With the fourth bid shaping up to be much more competitive than the third had been, the next challenge was setting IKN's bid price. Here again the structural tension between the Norwegian side (as official "strategic investors") and the Nepali SEL investors (who were putting up the vast bulk of the bid price) came to the fore. Hoftun agreed that Gyanendra Lal Pradhan, now officially general manager of the SEL group, should set the bid price, or at least "have the biggest say in the decision." But Hoftun, who had already found himself questioning G. L. Pradhan's judgement on several occasions, was also worried about Pradhan determining the price "all on his own." Hoftun hoped others within the SEL group would participate but he and Balaram Pradhan were also very concerned to limit the number of people involved and maintain strict secrecy—for fear that the other side would learn IKN's bid price and outbid them by a small margin. When asked for his advice on the bid price, Hoftun replied rather blandly, "It should be as low as possible, but not so low that we don't win the bid. I would say that the price ought also to be reasonable and high enough to ensure that HMGN is fairly but not fully satisfied." No one was to put IKN's bid price in writing, leaving Gyanendra Lal Pradhan in the "hot seat" as he walked into the bidding ceremony with the price only in his head.

In the end only two groups—IKN and IPCN—submitted bids but it would be months before anyone knew what those bids were. After the December 5 deadline the Privatization Cell had to form an "Evaluation Committee" and everyone knew from previous experience that the gears of government ground slowly. Ironically even Hoftun and the other Norwegians were left in the dark as to IKN's final bid price. Gyanendra Lal Pradhan and others within SEL were desperately concerned to keep IKN's bid a secret and out of the hands of the Chaudhary Group. They feared that if Binod Chaudhary learned IKN's price, "he may be able to change his envelope of financial bid through the help of MoF officials," Balaram Pradhan reported. A few days after the bid Hoftun wrote in frustration, "I am supposed to raise a lot of money for financing this bid. But SEL keeps things so secret that one should not think that we are partners."

That Hoftun was trying to raise Norwegian investment in BPC without even being privy to how much his own side had bid for the company only highlights the huge gulf that separated Nepali and foreign conditions in terms of business, politics, and simple security. From an international perspective Nepal seemed inscrutable at best and downright dangerous at worst. In fact even Nepalis were deeply worried. The weeks prior to the fourth bid had seen the collapse of Maoist peace talks and subsequent bombing campaigns around the country including in Syangja near the Andhi Khola plant. In one letter to Hoftun from early December 2001 Balaram Pradhan signed off with the exclamation, "NEPAL IS BURNING." Under these conditions Hoftun approached commercial investment insurance firms only to be informed that Nepal was "closed for risk coverage." In desperation he turned to Norad "to step in and provide guarantees for investors—considering the current situation in Nepal and the reluctance of Norwegian investors to come along." Reporting back to the SEL investors (who were still hoping that Norwegians would cover 65 percent of the BPC bid), Hoftun could only say that "the present situation in Nepal is not conducive for arranging fresh investments."

The year 2002 was full of incredible downs and ups for the IKN side, beginning with more unwelcome news. In early January SEL members reported having been approached by a Nepali Congress operative who invited IKN to contribute cash to the party in return for having IPCN disqualified from the bidding. Most "guessed that this proposal has been originated from MoWR Secretary," namely, Lokman Singh Karki. Gyanendra Lal Pradhan, fearing that his bid for the IKN side might be lower than IPCN's, thought that such a contribution might be a good investment. But others speculated that the Chaudhary Group had been given the same offer.

In Norway Hoftun continued his relentless efforts to raise investment funds but without great success. Though he had had some small successes (Agder Energi had agreed only to a disappointingly small NOK 3 million to be matched by Norad), by late January news came that all seven of the Norwegian utility companies that he had been aggressively courting for investment had rejected the prospect at the board level. By February Hoftun had only USD 1 million in committed cash but remained "very hopeful" of raising sufficient funds to cover the 30 percent deferred portion of the bid—a significant contribution though far short of a 51 percent controlling share, much less the 65 percent that the Nepali investors had hoped would come from the Norwegian side.

Finally word came down of a January 23, 2002, bid opening ceremony. To everyone's surprise, the Chaudhary Group's IPCN had bid exactly the same amount as in its previous three bids: NPR 866,000,001 or about USD 10 million. Also surprising was the discovery that Gyanendra Lal Pradhan had raised IKN's bid to NPR 874,200,000 plus USD 1 million cash for a total of NPR 959,250,000.[10] IKN had (seemingly) outbid the competition.

Although this was good news, the IKN side knew from experience that while they had won an important battle, the war was far from over. As expected, the Chaudhary Group immediately went to the press to sow confusion and accuse IKN of "irregularities." The day after the bid opening Hoftun wrote, "I had a feeling right away that Mr. Binod [Chaudhary] is again outsmarting us. Once I had read *Kathmandu Post* this morning my expectations were confirmed. He had his plan of creating controversy clear right away, ready to be implemented." Characteristically, the terms of payment for IPCN's bid were ambiguous. In its bid document IPCN

---

[10] A former NEA official I interviewed in 2017 insisted that the government of Nepal sold BPC "for nothing, very cheaply," basically NPR 1 billion for a total of 18.1 MW of power generation. He argued that the price should have been around USD 1,500 per kilowatt, which would have come to around USD 27 million or about NPR 3 billion. When I asked Odd Hoftun his opinion on whether the BPC sale price was too low, he replied, "Really yes, for someone who had any faith in the potential of BPC and its daughter companies." The problem was, Hoftun noted, that very few people had that faith. International investors certainly did not and Nepali investors were unwilling to risk such a large investment.

promised to make "payment within the final date of period" without specifying what period they were referring to. Rather than following the standard "terms of payment in accordance with the wording of the Bid Form," Binod Chaudhary "used double wording in order to get it both ways," Hoftun concluded. "If IKN was bidding higher, but with deferred payment, then he could pick the 100% cash alternative—as he has already done through the *Kathmandu Post* article. And if IKN's bid had been much lower, then he would, of course, have said that it meant deferred payment." The problem for the IKN side was that—because IKN's bid explicitly relied on deferred payment[11]—if IPCN did in fact pay up front, their bid might actually be higher than IKN's if interest rates were figured in. Even within the Privatization Cell officials disagreed on whose bid would be higher under various payment scenarios. If nothing else this meant that the lobbying war—to influence key decision makers—would need to continue full tilt.

By mid-February the government's evaluation committee still had not made a decision though Padma Jyoti reported that "signs look hopeful this time [even if] in Nepal you have to keep trying using all possible means until the last moment." In early March the government took one last dip into the BPC till when, at the BPC annual general meeting, officials announced a 10 percent dividend payable to the government itself. Hoftun accused the government of "stripping the company" of NPR 200 million, not to mention large amounts of tax. "And still they want the price to go up!"

Through the rest of March 2002 the emotional roller-coaster ride continued, beginning with the wonderful news on March 14 that IKN had finally been named "preferred bidder" and would be called for negotiations with the government toward finalizing a sales and purchase agreement or SPA. Immediately Hoftun wrote to Balaram Pradhan thanking him for "all your visits and talks everywhere during the last two or three years. I am so glad that you have stuck at it. Our cooperation has been extremely fruitful

---

[11] IKN's bid terms of payment were for 10 percent initial payment upon notification of being named preferred bidder, 60 percent payment within one month of signing the formal sales and purchase agreement (SPA) with the government, and the remaining 30 percent within two years of the signing.

and pleasant. And now comes the most exciting part!" Indeed Pradhan had invested phenomenal amounts of time and energy into the privatization process and IKN's bid. But, if anything, the coming "most exciting part" would require even more of his finesse and expertise.

Then on March 31 IKN's excited contentment came crashing down with news that Maoists had bombed and seriously damaged BPC's Jhimruk power plant. Already a few weeks earlier Maoists had cut the stay wires on several high-tension transmission towers near Jhimruk, causing them to bend and the wires to come dangerously close to each other. They then announced that they would not allow repairs unless they received an NPR 800,000 payment and the right to shut down transmission to Rolpa and Pyuthan on demand. When their demands were not met immediately the Maoists apparently decided to inflict much more serious damage. Bombs destroyed major components in the Jhimruk powerhouse. Though causing much less damage, Maoists also managed to bomb BPC's Andhi Khola facility, forcing its staff to flee and the power plant to go idle.

The Jhimruk bombing again sent the entire privatization process into a tailspin. IKN may have been the "preferred bidder" but no one even knew if the bid was still valid. IKN's bid was based on an intact Jhimruk plant. Was IKN obliged to make its required initial 10 percent payment to the government or was this now moot?[12] Should IKN withdraw their bid? Should the plant be repaired *before* negotiations proceeded? Who would pay for repairs? Who would cover the loss of Jhimruk revenues if privatization went ahead before the plant was back in running condition? In April, Hoftun—who had been in Nepal for a month during the bid opening, early negotiations, and the bombing—confessed that "after four weeks in Nepal I left quite confused, and am still feeling confused."

---

[12] Documents said that the initial payment should be within a week of IKN's official notification but it wasn't clear whether that week started when the Privatization Cell made its decision, or after the cabinet had met and confirmed the decision, making it legally binding and official. If the former, then IKN was already in default. If the latter, no one knew when such a decision would be forthcoming. Plus, after the bombing, was IKN obliged to pay for damaged goods? Or would the whole bidding process have to be repeated?

Confusion reigned as negotiations entered a period of stalemate. For the Nepal government, in the midst of a civil war, BPC's privatization was not a major priority. The government-run BPC board (chaired by the anti-privatization Lokman Singh Karki who was perpetually on junkets abroad) went months without meeting and was ineffectual when it did. For the Nepali SEL investors, the bombing paralyzed their efforts to acquire bank loans to cover their part of the BPC purchase price. No bank was going to lend money to buy damaged goods. The bombing also deeply undermined the SEL investors' confidence in the whole enterprise. Were they crazy to risk big money on such a volatile, vulnerable investment? And from an international perspective, the bombing only confirmed every foreign investor's worst fears about Nepal as a high-risk country. The bombing effectively destroyed any hope of raising further foreign investment. In the meantime Binod Chaudhary was lurking about, watching IKN's every move and, in Balaram Pradhan's words, "trying to fish in the troubled water." Chaudhary threatened the ministry of finance with a lawsuit if BPC went to IKN at a lower price than what he had, ostensibly, bid.

In fact no one really even knew exactly how badly Jhimruk had been damaged. BPC staff had fled in fear for their lives and, in the coming months, BPC management was afraid to even visit the site for assessment. And there was the question of who could be trusted to make an accurate, impartial estimate of what it would take to repair the power plant. IKN and the government agreed that it should be an independent third party, but who would pay the consultant? With the process in deadlock, Jhimruk sat empty and idle for months and months.

To break the impasse, in early May the Norwegian embassy sent an inspection team to Jhimruk, without Hoftun's knowledge, and issued a report. At a minimum they recommended repair to the powerhouse roof and windows to prevent further damage during the upcoming monsoon. But with the BPC board missing in action, and no one else in the BPC administration authorized to take independent action, nothing happened. An irate Hoftun wrote to Gyanendra Lal Pradhan:

> I think this is INCREDIBLY IRRESPONSIBLE. BPC's two power plants are both out of action. The company has no income, only expenses.

There is urgent need for at least to repair the roof and the windows of the powerhouse before the monsoon. BUT NOTHING IS BEING DONE BECAUSE THE BPC BOARD CHAIRMAN IS OUT OF THE COUNTRY FOR WEEKS and NOBODY ELSE CAN MAKE ANY DECISION! [*sic*]

BPC management was so paralyzed that they actually allowed the insurance policies on their power plants to expire, leaving the entire corporation extremely vulnerable to further attacks. Fearing the loss of what little foreign investment he had, Hoftun went to the government demanding "some sort of *pakka* [robust] assurance from HMGN's side regarding future security for the power plants" but got only verbal promises.

By early June 2002 both IKN and the government were still locked in a confused, muddled embrace with no one clear on how to go forward. The government accused IKN of stalling and made vague threats to back out of the privatization process if IKN did not make its 10 percent deposit payment. But IKN complained that it had never been formally or legally informed of its official "preferred bidder" status. Additionally, IKN refused to go forward until it had a formal written agreement with the government regarding how the Jhimruk repairs were to be taken care of and by whom. The easiest option would have been for the government to deduct the cost of repairs from the BPC bid price and be done with it. But this would have brought condemnation from those who already claimed that IKN's bid price was too low, and possibly lawsuits from Binod Chaudhary who had accused the government of favoritism toward IKN all along. Both sides waited for the other to make the first move.

For the IKN consortium the 10 percent initial payment was becoming an increasingly contentious topic. In earlier agreements the Nepali SEL investors had agreed to cover the initial 10 percent. But with so many elements of the bid in limbo, with the banks hesitant to lend money for damaged goods,[13] and

---

[13] In fact by late June 2002 the SEL investors had been able to secure bank loans for less than half of the sum they were committed to for financing the bid. Hoftun accused Gyanendra Lal Pradhan, who was leading the financing effort for the SEL side, of being uncommunicative if not secretive in refusing to keep the Norwegians up to date

terrified to throw their own personal funds into the seeming abyss, the SEL investors began asking Hoftun and the Norwegians to foot the bill for the 10 percent. In frustration Hoftun complained that the USD 1 million that the Norwegian side brought to the bid was unavailable. Part of it was already tied up in the bid bond guarantee and, what's more, Nepali law required that foreign investment be made as a lump sum, not split up as this maneuver would require. Pradhan explained to Hoftun that the SEL investors (not to mention the Nepali banks) would feel much more confident if the Norwegians were willing to put their money down first. "If IKN gives such support they will be willing to take further risks," said Pradhan. "Otherwise, they may lose patience and even start thinking of good excuses to back out." Tellingly, Hoftun shot back that SEL's refusal to pay even the 10 percent deposit would look very bad to the bid's few remaining foreign investors (Agder Energi and Norad), not to mention other potential investors. If the Nepali investors have so little faith in the project, how can they expect foreigners to join the effort?

The result was a fascinating catch-22 situation: the Nepali investors were scared to enter a new and unknown area of business; and the foreign investors were scared of Nepal's dismal socioeconomic condition. Each side wanted the other to take the first step as a vote of confidence. When Hoftun refused, as a last resort Pradhan asked for a "comfort letter" to the SEL investors reminding them that the Norwegians had determined that BPC was "financially viable and a good investment" and that IKN Norway would be "equally responsible to operate the Project with its partners" in Nepal. As a compromise, the Norwegians offered to make available USD 100,000 as a bank guarantee for money to be loaned in Nepal to SEL for the 10 percent payment.

In late June the Norwegian embassy again moved to try to break the logjam by offering to appoint (and pay) a specialist to act as mediator between IKN and the government. Quickly the ministry of finance agreed to issue an official letter (dated July 1, 2002) accepting IKN as preferred bidder with the stipulation that the 10 percent payment had to be made by July 8. Only after

---

on finances. Hoftun also complained that, if the banks were reluctant to loan, why not bring in other Nepali investors who had shown interest in joining the SEL group. "I don't understand this," Hoftun wrote. "Gyanendra wants to run the show alone, but then he must have the money to do so."

that would the SPA negotiations begin. But Hoftun was furious that IKN had still gotten nothing but verbal promises from the government. "What is HMGN going to do about Jhimruk?" he demanded to know, still trying to round up investment.

> We want those verbal promises put down on paper in clear text. Up to now we have only seen insults and broken promises.... How can investors and banks have any faith in HMGN this way? Norfund, for instance, will not give us any money to invest unless the Jhimruk issue is settled and nailed down on paper.

Hoftun's exasperated and rather harsh tone illustrates the difficult, tense, and high-stakes phase that the privatization process had entered in the final SPA negotiations. Again the structurally ambiguous question of leadership came to the fore: who—IKN Norway or the SEL investors—was going to make decisions and how? Hoftun knew that the SPA was a crucial document that would have important implications for BPC's future in general, and for laying out IKN Norway's role in a privatized BPC. For Hoftun any hopes of having majority control were long gone but he still wanted to make sure that IKN Norway's advisory role in the new company was clearly and formally established. For example, the SPA would spell out who had the right to appoint new leadership for BPC, a right that Hoftun wanted to hold. In fact it was this kind of influence over BPC's management and corporate vision that Hoftun, Pradhan, and others had fought for from the very beginning. But the SEL investors had their own interests to pursue and protect. The result was a period of heightened tension and fraying nerves.

IKN's negotiating team was basically Balaram Pradhan representing the minority Norwegian investors and Gyanendra Lal Pradhan representing the majority SEL group (with assistance from Ratna Sansar Shrestha, Padma Jyoti, and Rajiv Rajbhandari). But whereas the SEL representative was given rather extensive authorization to make decisions for his group, Balaram Pradhan had to report back to Hoftun who, in turn, had to get authorization from the full IKN Norway board before he could make decisions. The SEL investors had hoped that Hoftun would come to Nepal for the SPA negotiations but he could not.

Therefore it fell mainly to Balaram Pradhan to hold the chain of communication together through unreliable email, across continents and across languages. Stuck in the middle, Pradhan also became something of a lightning rod for Hoftun's frustration and anxiety. In addition to carrying the heaviest load in the negotiations, Pradhan was to be the eyes and ears for the absent Norwegian side. SPA talks with the legal department of the ministry of finance were tense with the government taking a rigid, condescending, "take it or leave it" approach that, Pradhan believed, Hoftun would have found so insulting that he would have left the room. (Pradhan credited Padma Jyoti with eventually using great tact to bring the sides together.) After long and trying hours at the ministry, Hoftun expected Pradhan to come home and compose detailed memos laying out the nuances of the day's negotiations. Because writing in English was tedious and time consuming, Pradhan tended to write more basic emails and then wait to elaborate on specific points when asked. Eager for information, Hoftun found this annoying and, at times, infuriating.

In one notable exchange from July 2002 Hoftun allowed himself to vent his frustrations onto Pradhan. With SPA negotiations underway and making good progress, the SEL investors decided to take a more active role to try to speed the process to an early conclusion. News of this more active leadership from the SEL side only stoked Hoftun's insecurity over the prospect of losing control of the SPA negotiations and, potentially, future influence over BPC. Hearing of SEL's plans to "speed up privatization," Hoftun was incredulous that they would try to do so without consulting IKN Norway, "who after all is the lead bidder! If SEL think that they can do this whole job without IKN's involvement, then perhaps IKN should pull out?" Hoftun was willing to let SEL (as majority investors) take leadership in some things, such as the bid price. But when it came to SPA talks that would determine IKN Norway's future role in BPC, Hoftun wanted to be in control.

No doubt frustrated, Hoftun took the occasion to take a nasty potshot at Balaram Pradhan, accusing him of aiding and abetting SEL who had, in this moment, become almost the enemy. "Sometimes I wonder whether you feel that you are employed by SEL, and only need inform IKN very briefly when Gyanendra ji tells you to do so," Hoftun wrote derisively. Pradhan was in the delicate position of needing to both promote the successful, and timely, privatization of BPC, *and* protect the interests of one side of an ostensibly unified consortium. When informed that the SEL side, in an effort to streamline the process, had decided not to register a joint holding company (to officially handle IKN's bid financing),[14] Hoftun found the news "even more surprising and shocking." In a sarcastic and accusing tone, he wrote to Pradhan, "Shall IKN just take note that SEL is thinking to scrap all old plans and agreements and tender regulations and all that we have based our bid on up to now? This is incredible. You owe me a long letter to explain what is behind this." Obviously Hoftun felt great pressure to hold the process together, keep the Europeans on board, and prevent another bid failure. But unable or unwilling to be in Nepal for negotiations, much of his anxiety got vented onto Pradhan. For Pradhan, these were some of the most trying and disappointing times in his long relationship with Hoftun.

This stress was almost certainly a contributing factor to a serious heart attack that Balaram Pradhan suffered in late August 2002. After receiving angioplasty and stents in a Kathmandu hospital, Pradhan returned home with an order for "complete rest" for the month of September.

In the meantime progress had been made on the matter of Jhimruk repairs. In an effort to break the impasse and hopefully spur on the privatization process, Norad agreed to provide NOK 22.2 million in grant funds to refurbish Jhimruk. But, they stipulated, these funds would only be handed over to a *privatized* BPC and the entire offer would be withdrawn if privatization had not been concluded by December 17. This move threw the ball into the government's court, forcing them to either take official action to wrap up the privatization or risk losing a major grant.

In October 2002 the Nepal government was again in tumult with a new prime minister and cabinet trying to gain footing in a chaotic social and political context.[15] In November a now somewhat-recovered Balaram Pradhan

---

[14] According to the bid documents, such a holding company was legally required. However, pointing to the fact that several previous government privatizations had ended successfully *without* such an arrangement, the SEL investors decided to ignore this requirement. In the end they were, in fact, correct.

[15] The unpopular King Gyanendra had sacked Sher Bahadur Deuba and installed Lokendra Bahadur Chand as prime minister.

reported that the whole government "is very weak and has not enough courage to take any such decisions which might raise question against it afterwards." Pradhan feared that in such a paralyzing and fear-ridden environment the whole privatization process would grind to a halt such that it "collapses by itself and there is no one to be blamed." "It is very easy to stop privatization," Pradhan observed, "but very difficult to get BPC privatized."

For the IKN–SEL negotiators, this was a period of incredible frustration. By October they had managed to hammer out a draft SPA with the ministry of finance but now the document had to go to the law ministry (under the new government) for legal vetting before it could be forwarded to the cabinet for final approval. Suddenly, issues that had apparently been nailed down in negotiations with one ministry were again floating about as the next ministry threw previous agreements out the window. Now anything that might appear to be a concession on the part of the government to the buyers became a risk for any officials involved. Pradhan reported that officials were being completely unreasonable and threatened to scuttle the whole affair on the slightest pretense.

At this point Pradhan was nearing the end of his rope. He was under enormous pressure to produce results, but also felt that he lacked bargaining power because he was not authorized to make decisions. Fearing a repeat of earlier unpleasant exchanges, and with the Norad deadline only days away, on December 14 Pradhan begged Hoftun to come to Nepal.

> I very strongly feel that you must try to come yourself or send someone from Norway, who can loudly laugh at them when they talk like this. You are trying to guide me from Norway, and I have not been able to handle the situation properly.... HMG is not negotiating with us in a proper way. HMG takes the decision and then asks us to agree the point. In case we do not agree, they threaten us to cancel the bid. Finally I want to express that BPC has reached to the point where it needs a very strong person (preferably WHITE FACE) for negotiation. [*sic*]

Otherwise, Pradhan feared, officials will take the path of least resistance or danger and simply put BPC up for retendering again. Pradhan did not want to be left to blame for another failure.

In response to Pradhan's plea Hoftun apologized for putting such a burden on the Nepal side but wrote that he himself was barely holding up to pressure coming from the IKN board and investors from the Norway side. Hoftun described a three-hour IKN board meeting at which the group came very close to throwing in the towel.

> To negotiate with HMGN does not work because there is no give and take. HMG decides and we are to listen and do whatever they say. And we must respond immediately, while Government can keep us waiting and guessing for weeks and months. On top of that, it appears that the Government changes positions frequently, throwing out old promises and commitments, and bringing up new demands.

Ultimately the IKN board decided to stay in the game but to stick to their principles and insist that proper procedure be followed.

In a final heroic flurry of meetings, lobbying, and transcontinental faxing, the IKN–SEL negotiators managed to get a "final draft SPA" to Hoftun on Tuesday, December 17 (the day of the Norad deadline) and begged him to sign it. Hoftun sent it back with corrections saying that he was not going to authorize its signing until he had a clean "final FINAL version of the SPA"! At this point Norwegian embassy officials, who had been closely following developments, agreed to extend the Norad deadline until December 19, but absolutely no later. On Wednesday, December 18 Ratna Sansar Shrestha combed through the SPA one last time while Gyanendra Lal Pradhan and Padma Jyoti visited every government official they could find before learning that the matter would go to the cabinet the following day.

Now at last the stars seemed to align in IKN's favor. The recently installed government had appointed Dipak Gyawali as the new minister of water resources. Unlike previous occupants of this position, Gyawali was an engineer by training with significant hydropower experience.[16] But more importantly, Gyawali was a strong advocate of private-sector power

---

[16] Gyawali had degrees in hydropower engineering from Moscow and political economics from UC Berkeley, along with decades of extensive first-hand experience with Nepal's hydropower development policies and debates.

production. To that end, he believed, BPC needed to be privatized, the sooner the better. "Here was an asset and it's going to be junk if we don't get it out of government hands," he told me in 2016. He also believed that Nepal needed competing power producers. "You can't only have NEA. The more the merrier." Nepal needed a private power sector and liberating BPC from government hands was a step in that direction. The new finance minister, Badri Prasad Shrestha, agreed that the time was ripe for getting rid of BPC and its damaged goods (the inoperable Jhimruk plant). "We can't keep a dead body on our hands. We've got to get rid of it before it starts festering and stinking."[17]

These two key ministers were in favor of BPC's privatization but they faced huge resistance from within their own ministries and from the shaky coalition government as a whole. They could override their underlings but they could still be accused of mismanaging or underselling government assets (like BPC) or, worse, of personally profiting from the transaction through graft. There could be political and even personal fallout from their decisions. Nevertheless, Gyawali and the "very nervous" finance minister Shrestha officially approved the BPC SPA and arranged to have the matter brought before the full cabinet on Thursday, December 19. Gyawali used the Norwegian Jhimruk grant deadline "as a weapon. And that was partly the reason for success."

As Gyawali recounted the story to me, the cabinet at the time was a mixture of various parties including "three hardcore communists" who were "so allergic to the very word 'privatization' that we knew it was going to be a problem." The finance minister felt that the BPC proposal had no chance of getting past these cabinet members but Gyawali had a plan. According to Nepal government rules, documents to be brought before the cabinet for discussion were to be delivered to the members' offices or residences at least

---

[17] In our conversation Gyawali implied that, by late 2002, opposition to BPC's privatization was diminishing even from within the Nepal Electricity Authority (NEA). With its power plants bombed and out of operation, BPC was no longer an active moneymaker. Said Gyawali, "NEA isn't interested in running things where they haven't got contracts and commissions and kick-backs and things like that. So they ran it but it was being run down and then the Maoists came in and bombed Jhimruk."

two days in advance. But, said Gyawali, this rule is "flouted every time. It always comes the evening before, if at all."

> So what I told Badriji [Badri Prasad Shrestha, the finance minister] is, "Listen, let's not leak this out first. If we send this out two days in advance, it's going to cause problems. The news will be all over the place and these guys will get pressure from supporters not to approve it."

If the BPC matter was sprung on them, the anti-privatization interests would not be able to mobilize against it. Gyawali noted that if the cabinet was split, the matter would be sent to a subcommittee where it would flounder and die.

Gyawali had the BPC SPA documents delivered to cabinet minsters' homes at 8:00 p.m. the night before the meeting, sure that "they'd be out at a party and too tired to read it when they got home." The next morning when people realized what was going on, there was resistance. But Gyawali noted that the Norwegian grant money was in jeopardy if they did not act immediately. "I said, 'If you don't do it, the grant will expire. Are you prepared to let this grant lapse?' Ahhh, there was rancorous debate!" Gyawali told the other cabinet members that

> this is just about the best deal we can get. If you get any flack for this, you can blame it all on me. So very reluctantly they said, "OK, but if anything goes wrong, we're blaming you!" So I said "OK," and that's how it got passed! And then it all went through without a rumble, without any negative press, without anything. It was all so transparent that there was nothing to complain about [even though corruption investigators were paying close attention].

With the privatization now formally approved by the government, the following day (December 20, 2002) Balaram Pradhan signed the official agreement with Norad authorizing the transfer of funds to BPC which was to take full responsibility for the Jhimruk power plant's rehabilitation.[18]

---

[18] The NOK 22.2 million grant stipulated that the Jhimruk repairs had to be completed within fifteen months and within budget. For its part, the Nepal government agreed

According to the final SPA the Nepal government agreed to sell 75 percent of its shares (6,292,932 out of 8,390,755 shares) in BPC to the IKN consortium. (Of the remaining 25 percent, 10 percent were to be sold to BPC employees and the public with the rest held by the United Mission to Nepal [UMN], the Nepal Electricity Authority [NEA], and the Nepal Industrial Development Corporation [NIDC].) The final sales price was NPR 950,880,000 or USD 12,400,626 at the exchange rate in 2002. Of the total bid price the SEL investors provided NPR 874,200,000 for 5,785,463 shares and IKN Norway NPR 76,680,000 (USD 1 million) for 507,469 shares. The agreement stipulated a seven-person board of which two would be government appointees and five appointed by the purchaser.

By the end of December 2002 the privatization process had cleared the cabinet approval hurdle, and secured the Norad grant, but the SPA still was not formally signed—apparently to give IKN time to begin mobilizing funds. Once the SPA was signed, IKN had 30 days in which to pay 60 percent of its bid price. In fact the SPA remained a work in progress. Because finance ministry officials insisted that the "official bidder" (IKN Norway) also be the official purchaser, the government had removed from the SPA all mention of the SEL investors and their role in financing the bid. This was ironic to say the least given that, while the Norwegians were committed to covering only a small portion of the bid price, the SPA held them to be legally responsible for fulfilling the terms of the contract. In frustration Hoftun wrote to Padma Jyoti, complaining how "here I am in the end left responsible for this IKN that has been turned into a single Purchaser carrying the total responsibility for this deal without being able to have much influence on what sort of terms and conditions we now will have to abide with." In the final weeks of 2002 Hoftun tried desperately to get wording into the SPA that would officially acknowledge that IKN Norway and SEL were jointly undertaking the bid and would jointly accept its obligations. But with little leverage, Hoftun's requests were ignored, leaving him fearful of being left holding the bag.

Tinkering continued until January 3, 2003. Even on the last day there were hours of "very heavy discussion" with ministry officials looking for any

---

to forego all taxes, fees, and duties on all equipment and materials needed for the rehabilitation work.

anomaly that would justify their scuttling, or at least postponing, the process. But, accompanied by legal experts and bank officials, the IKN–SEL team finally forced the government's hand and got the SPA signed. "The main part of the Privatization of BPC is completed today at 16:00 hours," Pradhan wrote to Hoftun.

This was welcome news for those BPC employees who had stuck with the company through its years under government control. Pradhan reported that "BPC's present employees are favorably inclined toward private ownership ... because they have been used to it in the past and are confident that the new private owners can offer good management practices." They looked forward to a return to the dynamic leadership and growth that the company had experienced in previous decades.

Looking back at more than half a decade of work on privatization, and many decades of collaborative work, Pradhan wrote to Hoftun, "I am really glad that BPC privatization has been successful after so long a time. It has been like one more Project for us together which now has been successful. It was not simpler than building Jhimruk Project." In a thirty-five-year career of long, complicated "Projects" working together with Hoftun, for Pradhan this was the most difficult one. Compared with dealing with the government of Nepal, building a major hydroelectric facility from scratch was easy! In response Hoftun expressed his deep thanks and noted, "You have worked on this project more than anybody else." "I really need a few weeks of holiday," concluded Pradhan.

But still to deliver was payment of the 60 percent of the bid fee due thirty days after the SPA's signing. The SEL investors managed to secure loans from a consortium led by Himalayan Bank Ltd. (on whose board one of the SEL group was a member). And while Pradhan took a well-deserved break, Hoftun continued full speed for three more hectic weeks arranging the complicated legal documentation needed to transfer money from Norway to Nepal. Finally, after an eleven-hour marathon session in Oslo on January 27 ("There was excitement right up to the end!"), representatives from IKN, Agder Energi, and Norfund signed documents releasing the investment funds.

The official transfer of BPC occurred on January 29, 2003. That morning the BPC board met one last time under government management and then met again in the afternoon as a privatized entity. At BPC headquarters the government officially transferred its shares into IKN–SEL's hands. It was a festive event, made even more so by unplanned music provided by a nearby wedding party band! Present were old and new BPC staff and board members, ministry and Privatization Cell officials, SEL investors, Norwegian embassy officials, and other well-wishers. At its second meeting of the day the BPC board approved the appointment of Egil Hagen as general manager, and Suman Basnet as deputy general manager. Both Hagen and Basnet were veterans of earlier BPC projects, handpicked by Hoftun to lead the new BPC (as per conditions set down in the SPA).

Absent from the BPC management team was Balaram Pradhan, even though Hoftun had been courting him for the general manager position literally for years. But Pradhan firmly refused. In January he wrote to Hoftun that the previous six years had "proven to be very expensive to my health…. I want to stop working for IKN as quickly as possible" and "try to keep away from BPC matters." Both Egil Hagen (the new general manager) and Gyanendra Lal Pradhan (Balaram's uncle and leader of the SEL investors) had offered him positions as paid advisor, but he refused saying that he wanted nothing to do with more meetings and decision-making.

Balaram Pradhan's decision to extricate himself from BPC matters after so many years was surely a matter of health concerns and simple exhaustion. But it is also clear that, perhaps more than anyone else, he understood the potential for conflict that was embedded in the new BPC's leadership structure. In short, the same ambiguities and leadership contradictions that had plagued the entire privatization process were almost sure to come to a head in the now privatized company. Already a battle was shaping up between Hoftun and the IKN Norway board that wanted to maintain managerial and moral control over the company, and the SEL investors—and especially their most outspoken and aggressive member, Gyanendra Lal Pradhan. Hoftun wanted to return BPC to its earlier days as a leader in human capacity building and ethical growth. But the SEL investors, now deeply in debt, needed BPC to be a profit-generating enterprise. They had invested in a business, not a social service.

Balaram Pradhan sensed an inevitable clash. A year earlier, when Hoftun has asked him to join the BPC management team, Pradhan wrote of being "puzzled" and dismayed by the pullout of Norwegian investors. Whether

Norwegian commercial majority investment would have guaranteed the smooth implementation of Hoftun's management vision for BPC is anyone's guess (though it seems likely that this combination would have produced the same clash between commercial and service motivations). But Pradhan knew that, under Nepali commercial ownership, it would fall largely to him to bridge the gap between Hoftun and the SEL investors, trying to keep the two conflicting drives from ripping apart. "Now with Nepali majority investment it might bring many tough situations," Pradhan wrote presciently. "I feel that I should keep myself away from the Management team."

In fact already for several years Pradhan had been looking into alternative employment options, after BPC privatization. Even while privatization was underway Pradhan had accepted a part-time position with Wasserkraft Volk (WKV), a German hydropower equipment manufacturer, as their official representative in Nepal where WKV wished to develop hydro projects. With privatization now completed, Pradhan planned to take up this position which he hoped would be much less stressful.

With the transfer at last completed, at the end of January 2003 Pradhan wrote to Hoftun looking to formally wrap up the project. "So let us agree that we have completed our JOINT MISSION. I would appreciate a Completion Certificate from IKN." But after years of rapid correspondence, Pradhan waited almost three months for a reply. Hoftun wrote at last,

> Well Balaram, I have been under lots of pressure and felt tired and unable to get my work done in an efficient manner. The closing chapter of the BPC privatization process really sucked the energy out of me. When finally the word came through that the payment happened in time and the BPC sale shares had been turned over, then I felt exhausted and fed up with it all. With my 75 years of age I don't have much energy in reserve. Everything takes longer time to do than it did in the past. Even to write an email like this feels like a big job.

Again Pradhan asked for a letter of reference.

> I think it is good to have one letter from you as a souvenir. I have got such letters from you several times and I am used to receiving such letters from

you which I keep for my own purpose as "Souvenirs from the Past." Maybe that will be useful for some good cause? While you send the letter please send me your latest photograph as well. I want to keep your photo in my office.

Pradhan was in a sentimental, reflective mood looking back on a long relationship with Hoftun: a man he saw as an important mentor and even father figure; a man whose values and motivations closely aligned with his own and, in some ways, had helped to shape them; a man whose career had been closely intertwined with his own.

Therefore, it is hardly surprising that Pradhan felt confused and upset when, a year later, he still had received no reply from Hoftun. Finally, in an email from April 2004 (headed "After a very long time"), Hoftun wrote back saying that he had heard indirectly that Pradhan "felt hurt" and was wondering if Hoftun "had something against him." "I now want to make it very clear that I have no ill feelings of any sort against you," said Hoftun. He blamed the lapse on "being overloaded with work" and old age. He wanted to write a long and meaningful letter, but the thought of the effort involved made him put it off.

No doubt Hoftun was being sincere but his reference to "being overloaded with work" points to the fact that the end of the privatization saga was anything but the end of Hoftun's involvement with BPC. Pradhan may have tried to pull out of the BPC drama but, a year after privatization, Hoftun was back in the thick of it. And, in fact, in his new role with WKV, Pradhan had managed to raise Hoftun's ire in ways that he certainly never intended.

8

THE NEW BPC

Cultures in Conflict

In 2003 the now privatized Butwal Power Company (BPC) set out to retain and expand its leading role in hydel development in Nepal. The company's new joint owners—the Nepali majority investors who made up the Shangri-La Energy Limited (SEL) group and the minority Norwegian investors of Interkraft Nepal (IKN)—were eager to make up for lost time and institute their visions for the new company. Unfortunately, these visions were often not eye to eye. Whereas Odd Hoftun and a number of Nepali professional staff with ties to the old BPC sought to re-establish the company's earlier ethic of corporate reinvestment and national capacity building, the new Nepali majority owners were businesspeople intent on generating profits—not least in order to service debts incurred in buying BPC. Whose vision would prevail became a matter of pointed struggle as IKN tried to assert its rights to management control (legally provisioned in BPC's sales and purchase, and consortium agreements) while the SEL investors claimed their rights (and interests) as majority investors. The structural power fault lines already under strain during the privatization process only increased their friction as the two sides repeatedly rubbed against each other before finally splitting and drifting apart. Over the following decade the collateral damage included a chronically underperforming BPC that saw its leading status ebb away as new independent power producers entered the market, and an increasingly dispirited Hoftun who watched the gradual demise of the development philosophy that had motivated his involvement in BPC for most of his life.

POST-PRIVATIZATION: THE EARLY YEARS

According to the management agreement arrived at earlier between IKN, SEL, and Agder Energi (the Norwegian power company that was to play an advisory role in the now privatized BPC), Agder Energi would make available a "full time expatriate expert to fill the position of BPC General Manager." This person was expected to train and eventually hand over the position to a Nepali counterpart. The same agreement gave IKN the right to choose a Nepali national for the deputy general manager position with both manager and deputy formally appointed by the BPC board. For years Odd Hoftun had hoped that Balaram Pradhan would accept the general manager post but when he declined, IKN turned to the team of Egil Hagen and Suman Basnet. Hagen and Basnet were old friends, having known each other since the 1980s when both worked as electrical engineers with BPC at Andhi Khola. After working together for years Hagen had returned to Norway where he worked for a consulting firm and Basnet went to the United States on a Fulbright fellowship to earn an MBA. Together they brought an impressive array of technical and managerial experience, understanding of Nepal, and a deep commitment to Hoftun's development philosophy.

Perhaps mirroring the larger structural tensions built into the new BPC, one problem from the beginning was the nature of Egil Hagen's appointment. Though he had spent his early career in Nepal working at missionary wages, when he returned to his native Norway in the 1990s Hagen needed to make a Norwegian wage to support a family in a high-cost economy. When Hoftun asked him to serve as BPC's new general manager, Hagen agreed but on the condition that he would not have to break off his Norwegian career for what was, by definition, a relatively short-term position in Nepal.

Therefore, when Hagen and his family returned to Nepal, they did so on a full Norwegian salary with all the normal European benefits (for example, lengthy paid vacations) plus some extra perks associated with a foreign posting (housing, travel, and so on). A salary package that seemed reasonable for a European professional seconded to an Asian corporation seemed exorbitant to many in the new BPC—especially given that it was BPC that would have to foot most of the bill. Although Hagen was officially on contract from Agder Energi via a Norwegian personnel management firm, ultimately Agder Energi and IKN billed BPC for Hagen's services, among other things. Within a company struggling to find its financial footing and pay off debts, Hagen's salary was a source of resentment even if these expenses had been explicitly built into BPC's long-range business plans prior to privatization. For those looking to undermine Hagen's leadership in the new BPC, salary was a convenient criticism. In fact, even many who supported Hagen acknowledged that his compensation seemed excessive, especially in the Nepali context.

Later in this chapter I will return to look in detail at two major issues around which conflict focused (repairing the damaged Jhimruk power plant and "up-grading" the Andhi Khola project) but I want to begin by tracing the management struggles that characterized the first several years of BPC's post-privatization life.

From his earliest days of involvement with corporations in Nepal, Odd Hoftun knew that the ability to control a company's mission and destiny rested in those with the most power. In fact, the corporate model had appealed to Hoftun as an approach to development in a country like Nepal precisely because of the institution of the corporate board of directors. A board might include representatives of various stakeholders and interested parties (for example, the Nepal government) but as long as Hoftun and other like-minded persons could control a board majority, they could control the company. For that reason Hoftun had worked incredibly hard to recruit Norwegian commercial investors to buy a majority stake (and therefore board control) in a privatized BPC on the assumption or hope that this control would protect BPC from the "corrupting" influences of Nepali political and commercial culture. As that dream of majority or board control dissolved over the course of endless privatization machinations, Hoftun again worked incredibly hard to at least guarantee IKN control of key managerial positions through agreements written into the official privatization documents. In an April 2003 email to his old friend Balaram Pradhan, Hoftun expressed regret over having lost board control of BPC but took consolation in the fact that IKN could, through its placement of "first class people to fill key roles there," help guide BPC's "day to day management. This is real influence because it deals with realities!"

But in the same note Hoftun also expressed deep frustration. Just three months into the new dispensation there were already signs of serious problems. "I must confess that the latest developments in the relationship between IKN and SEL have been very discouraging.... There were signs of this earlier, as you are well aware." Succinctly, Hoftun concluded that "the question is how well the management people are able to work together with the owners through the decision making process." Given that, in the world of business, "management" is supposed to *represent* "the owners," here Hoftun acknowledges a fundamental and ultimately fatal flaw in the new BPC's power structure. Ultimately, would corporate control lie in the hands of "management" (IKN) or with "the owners" (SEL)?

Obviously these two were not inherently opposed: both the IKN-appointed management and the SEL owners wanted a prosperous BPC. But what remained ambiguous was the new BPC's chain of command and exactly where ultimate authority lay. It was this structural ambiguity, compounded by personality clashes and conflicting priorities, that eventually led to a showdown between management (IKN) and ownership (SEL).

Representing the two sides of this structural fault line were Egil Hagen and Gyanendra Lal Pradhan. Pradhan (along with his nephew Balaram Pradhan as well as Hoftun, Suman Basnet, Ratna Sansar Shrestha, and Dan Jantzen) had played a pivotal role in the privatization process. From the beginning he had been the de facto leader of the SEL investors and had worked extremely hard to secure the bid. As discussed in the previous chapter, it had fallen to Gyanendra Lal Pradhan to determine the final bid price—which he set at a higher rate than most expected. Pradhan could take credit for a successful bid but he also felt deeply obligated to his fellow SEL investors to make sure that the new BPC was profitable, not least because he and the others had to pay off deep bank debts with loan periods of just seven years. Gyanendra Lal Pradhan had professional training as an engineer and had provided leadership so far. Therefore, the other SEL partners were willing to delegate management authority to Pradhan who wanted to turn BPC into a moneymaker, and fast. Within weeks of the new BPC's launch, Gyanendra Lal Pradhan was frequently at Egil Hagen's door with suggestions and advice on how to manage the company. Feeling that his advice was unheeded, and frustrated by what he saw as a lack of progress, Pradhan also soon became Egil Hagen's most vocal critic.

For his part, Egil Hagen complained of being pulled in two directions without a clear chain of command. Responding to growing criticism from Gyanendra Lal Pradhan, in frustration Hagen wrote to BPC Board Chairman Padma Jyoti with the basic question: "Who do I answer to and take orders from? If I was to give to Gyanendra everything he asks for, that will take so much of my time that other important BPC work would suffer." Hagen understood his bosses to be the BPC board and therefore complained that Gyanendra Lal Pradhan had set himself up as "some kind of Managing Director" seemingly empowered to command the general manager (Hagen) and report directly to the board. This situation of ambiguous authority, Hagen complained, "may prove to be more than I can handle."

In fact the BPC board itself quickly embodied the same ambiguity and dysfunction that were playing out at the managerial level. Among other things, the various privatization agreements had established that IKN would have one seat on the BPC board and that a Consortium Executive Committee would mediate between SEL, IKN, and the BPC board. For IKN's board representative, Hoftun chose Ratna Sansar Shrestha, the high-minded, cantankerous, and unyielding chartered accountant and corporate lawyer who had aided BPC in the past and helped guide it through legal shoals during the privatization process. Hoftun and Shrestha viewed each other as fellow fighters in the cause of rooting out corruption in Nepal down to even the hint of ethical impropriety. Already in February 2003 Shrestha warned Hoftun that "I cannot suffer foolish things very graciously" and quickly established himself in an adversarial role on the BPC board where he adopted what to others seemed to be a nit-picking, obstructionist tone. Other board members quickly grew frustrated with Shrestha's behavior, seeing in him someone more interested in dissent than problem-solving. They accused him of refusing to take a productive, proactive stance toward BPC's many challenges but rather seeking to obstruct and delay even minor operational decisions. Viewing his job to be keeping BPC on the straight-and-narrow path, Shrestha adopted this combative tone from the very first new BPC board meeting, and relations deteriorated from there.

In theory the points of conflict between IKN and SEL views and priorities that Ratna Sansar Shrestha focused on during BPC board meetings *should have been* at least partially resolved prior to meetings by a Consortium Executive Committee whose constitution had been agreed upon by SEL and IKN during the privatization process. According to the Consortium Agreement that formed the basis for SEL and IKN's joint bid, an Executive Committee consisting of two representatives from each side should have been established immediately to hash through issues so as to propose unified advice to the BPC board before its monthly meetings.

But it was soon apparent that the SEL side, and especially Gyanendra Lal Pradhan, was none too keen on actually implementing this or other parts of the consortium agreement. From SEL's perspective, it was already a rather generous concession to allow IKN a seat on the BPC board, given that IKN represented only about 6 percent of the investment shares. Having IKN control *half* of an executive committee seemed even more egregious. For their part, IKN felt fully justified in trying to take a strong leadership stance because, as the official bidders and sole signatories of the sales and purchase agreement (SPA) with the Nepal government, IKN felt legally obligated (and eager) to fulfill the business plan laid out in the official documents (Hoftun 2004). But with privatization now a done deal, and the Nepal government's attention fully diverted by other pressing matters (such as a raging civil war), the SEL partners, and especially Gyanendra Lal Pradhan, felt no real compulsion to follow the letter of the law. For the Nepal government, BPC's privatization was water under the bridge and the SEL investors knew that no one would be looking to enforce the fine points of the SPA, especially if it meant siding with foreign minority investors.

Another problem with a Consortium Executive Committee from the SEL investors' point of view was the question of who would represent IKN on such a committee. IKN had appointed Ratna Sansar Shrestha—a man who was already a thorn in their side. But who would be the second IKN representative? At least Shrestha was an insider. The second member would have to be someone from outside, unlikely to ever really understand the fine points under debate. Eventually Hoftun convinced his old friend K. B. Rokaya (a well-respected public figure and human rights campaigner with a degree in engineering) to represent IKN on the BPC Executive Committee (and act as an alternative board member when Shrestha could not attend).

The Executive Committee managed to meet sporadically during the spring and summer of 2003 (with Rajib Rajbhandari and Pradhan representing SEL and Shrestha and Rokaya IKN) before eventually collapsing in acrimony and finger-pointing. Rokaya wrote to Hoftun expressing his frustration and confusion. In meetings he complained of feeling "uncomfortable to speak freely.... Gyanendra does not seem to recognize or consider me as a CEC member," for example, refusing to copy Rokaya on communications regarding committee business. It was not hard for Rokaya to put his finger squarely on the problem.

> I definitely see a strong conflict between IKN and SEL which keeps on surfacing in various forms. This situation should not be allowed to continue. In my understanding, *it is the tension between two cultures: the culture of BPC before privatization and the purely business/commercial/profit-oriented culture of SEL.* Conflict is not necessarily bad. It is just a question of how we manage the conflict. If properly and wisely managed, conflict can be good and can promote growth and progress. (Italics added)

To their credit, both sides recognized that, in Royaka's words, "this situation should not be allowed to continue." For IKN's part, Hoftun tried to reign in the overly zealous Ratna Sansar Shrestha. Already in April 2003 Hoftun delivered the first of several admonishments to Shrestha, pleading with him to take a more conciliatory approach to his BPC board duties. "You are a very principled person," wrote Hoftun. "When you think you are right, then you will stand for it to the end. That is good, and that is why I trust you very much."

> But at the same time you have a tendency to over-argue your case after you have made your point very clear. Sometimes you seem to lack the feeling of when the time has come to stop, when you ought to be a bit more accommodating in order to avoid unnecessary damage to relationships.

I wonder whether you might be at such a point? Please do not mind my frankness!

Honesty and forthrightness are good, implied Hoftun, but self-righteousness is corrosive and counterproductive. In May 2003 BPC Board Chairman Padma Jyoti wrote to Shrestha suggesting that he (Jyoti) might meet with Shrestha, Pradhan, and a few others with the aim of patching up relations. To this Shrestha was evasive, making himself difficult to pin down for a meeting and eventually responding that "I am not aware that I have any personal things to discuss with you all." In effect, Shrestha greeted Jyoti's olive branch with a disinterested shrug.

In June 2003 Gyanendra Lal Pradhan and Rajib Rajbhandari flew to Oslo to meet with Hoftun and the IKN board. The IKN board itself was increasingly alarmed by the deteriorating state of IKN–SEL relations and welcomed the two Nepali representatives. According to Hoftun, meetings and extended discussions included some "frank exchanges" that "helped clear the air," but little actually changed. In the midst of quarrels Hoftun administered some tongue lashing that the two visitors endured "because I [Hoftun] am the older person!"

In an interview with me, Rajib Rajbhandari reflected on a marathon debate that he held with Hoftun on a three-day tour of power plants in southern Norway during their visit. "Yes we had a lot of arguments. I admire [Hoftun] for his tenacity and his vision. There's no doubt about that. And his dedication and ideals. But I'll tell you, we had an argument for *three days*! For three days we had an argument about the philosophy of business."

> He and I disagreed on one point. He was saying that business must be so socially minded. My philosophy was: business needs to be business. With that income that you get from business, if you are socially minded you can do social things. If you are business minded you can grow. But you cannot let that govern the way you do business because then that would be a compromise. And that wouldn't be good for business.

"The business should be ethically run and everything," Rajbhandari continued, "but once the money comes into the business, you have more money to do good—if you are so inclined. But *the business should not be a charity!* The charity should be an outflow *from* the business."

> We had this argument for three days and we literally got thrown out of a train station! I'll never forget that. We were in this town called Kristiansand and we were supposed to take the train back. The argument was going on and on and was getting kind of heated. And for Norway—being so quiet, so gentle—our conversation was very loud. So the conductor who was there came over and said, "Please keep it quiet. People are very worried!" And Odd being Odd and me being me, this just went on and on and on. Even as we said goodbye, we were still arguing.

Rajbhandari concluded by noting that his family has for generations engaged in social philanthropy in Nepal. "But if you mix business and charity together, you can't do charity and you can't do business either. For us, doing business gives us a certain amount of money to be able to do charity."

In the epic struggle between what K. B. Rokaya had identified as "two cultures," neither side was much interested in giving up ground. But, from his relatively weak position, Hoftun tried to find a silver lining in this cloud of disagreement. In an email to Ratna Sansar Shrestha, Hoftun wrote:

> There may be a big gap between their ways of doing business and our ways. They are aggressive and rough and arrogant, but they are successful businessmen who have taken big risks because they believe in the potential of BPC and NHE. This is what we were seeking while we were fighting so hard and so long for privatization for BPC. [Still], IKN cannot simply walk out of the whole thing.

BPC's new owners may not share his development philosophy, but at least they are successful Nepalis committed to the company's success. IKN may ultimately have to fold to SEL's more powerful hand, but in the meantime, IKN cannot just give up. For its part IKN too was deep in debt, especially for a small nonprofit entity. IKN would have to stay in the game for at least as long as it had debt obligations to fulfil.

By the summer of 2003 Hoftun and the IKN board still held out hope that conditions would improve in Kathmandu once Egil Hagen hit his stride and "really starts functioning as the GM of BPC." But eventually even the Norwegians began losing patience over the slow rate of progress in BPC's management. Reports came in suggesting a lack of effective communication between Hagen and the board. In retrospect, Hoftun remarked:

> I think there was some weakness on his side in keeping the board fully informed or doing what is always necessary in Nepal. You talk to the individual members of the board and explain and try to get them to understand the reasons for proposing so and so. I see that he was too weak in communications.

Hagen's communication skills were not only somewhat weak but, at times, taken to be offensive. In one board meeting exchange that was described to me by several observers, Hagen is reported to have responded to a major proposal by (board member) Gyanendra Lal Pradhan by blurting out: "That is sheer nonsense!"[1] Even board members who were sympathetic to Hagen's views took offense at his tone, seeing it as a condescending insult to Nepalis. Hagen's friend and deputy general manager, Suman Basnet chalked it up partially to differing styles (European directness versus Nepali indirectness) but in the end even Basnet had only very faint praise for Hagen's performance.[2]

By late 2003 the IKN–SEL consortium was showing further signs of strain. In a nutshell, the SEL investors complained that their relationship with IKN and Agder Energi was costing more than it was worth. Even though these payments (which included Egil Hagen's salary) were agreed upon in the Consortium Agreement (and partially funded by Norad), like other parts of the agreement, the SEL side felt no real compulsion to hold

---

[1] The content of Pradhan's proposal is discussed later.
[2] In an interview with Dan Jantzen in March 2004, Basnet rated Hagen's performance as "maybe 5 or 6 on a scale of 1 to 10."

up their end of the deal.[3] Business service fees that were standard in Norway seemed exorbitant in Nepal, especially when Agder added a special "war risk surcharge" to their BPC bills (payable in foreign currency). In August 2003 when Hoftun had tried to get BPC board representatives to sit down with IKN and Agder Energi to discuss details of future "technology transfer," he found apathy on both sides. The fact that BPC was already delinquent on payments made Agder less than enthusiastic. And the SEL side had decided that it could live quite happily without Agder's services, especially if it meant saving money. Frustrated, Hoftun and IKN were left holding the bag. Technology transfer had been one of the key components of the formal business plan that the IKN–SEL investors had put forward in their tender documents and Hoftun felt that this technical assistance was one of the most valuable services that IKN could offer the new BPC in their stated goal of growth and development. But the SEL majority in BPC did not seem much interested.

From this point onward Gyanendra Lal Pradhan, de facto leader and representative of the SEL investors, was in open defiance of SEL's consortium agreements with IKN and maneuvering toward an administrative coup that would knock IKN out of the management picture altogether. Pradhan began publicly blaming Egil Hagen for BPC's lackluster performance since privatization and pressuring Hagen to report to *him*, not to the board at large. In desperation, in November 2003 Hoftun and the IKN board managed to get the BPC board to approve a "management contract" that clearly laid out Egil Hagen's role as BPC's general manager and the chain of command within which he worked, especially with regard to the BPC board. But within weeks of the new contract's signing, Gyanendra Lal Pradhan was openly calling for Egil Hagen's resignation. On the BPC board, Ratna Sansar Shrestha was pushed to the margins (accused of obstructionism) where his role was reduced to having his official votes of dissent noted in the minutes and, otherwise, being largely ignored.

---

[3] In an interview with me, Ratna Sansar Shrestha conceded that adhering to the Consortium Agreement was not required by Nepali law. When the SEL investors backed away from the agreement, they were violating a private agreement between shareholding parties but not violating Nepali law.

In the midst of this struggle, Odd Hoftun announced his resignation from the IKN board, having finally found a willing successor as board chairman in Mr. Tor Einar Ravnevand, a Norwegian lawyer. "I am happy to see an end to this long ordeal, which has become an increasingly heavy load for me," he wrote to Ratna Sansar Shrestha. "I feel that my capacity for this kind of work gradually has been reduced a good deal. And there are also other things I would like to get time to do before I die!" Yet, characteristically, within a week of bidding each other adieu, Hoftun was back in touch with Shrestha explaining that "although Tor Einar has taken over my responsibilities … I am still involved somewhat by advising and informing him when necessary until he gets on top of his job in every detail." After requesting to continue to be copied on all BPC correspondence, Hoftun told Shrestha that he viewed his ongoing role to "basically be as personal sparring partner for Egil. In order to function in that capacity I need to be as well informed about BPC affairs as possible." As he had before, Hoftun announced his retirement from the fight only to almost immediately step back into the breach.

Hoftun may have been incapable of withdrawing from the struggle between what Rokaya had called "two cultures" within BPC—the pre-privatization, "socially minded," Hoftun-oriented culture and the new post-privatization SEL business culture—but other holdovers from the old BPC were already voting with their feet and leaving. One Nepali engineer I spoke with, a veteran of the Jhimruk and Khimti projects, described how he had returned to the new BPC only to leave almost immediately. "When I came back I worked for one month in the same department where I used to work, then there was a kind of negotiation between me and the company. But we couldn't agree on the terms and conditions. So I left." When I asked what had driven his decision to leave, he explained:

> It was a combination of a few things. After the privatization, the owners of BPC, it seemed as if they wanted to run the organization as their personal private company. But our habit had been to be an employee of a broad corporate *institution*. But instead, there was the feeling of being employed by a small company that is owned by a single person. So that in a way discouraged me. After privatization, BPC was owned by business people.

I heard this same sense of disappointment—of no longer working collectively for some larger societal goal but rather toiling for some owner's personal profit—from other old BPC employees who left the company in 2003. Interestingly, several of them, like the person quoted earlier, used the word "institution" to describe the old BPC. "I would say that BPC [previously] was not a company but an institution," one explained to me.

> As an institution it was like a university for Nepal. It should have continued in that way. Because training people is very important. One or two clever people cannot do all the business. So there should be mass movement. It is about teamwork. When BPC became a profit center, profit oriented, it could no longer serve this function.

This person understood clearly that the old BPC's "socially minded" mission was not about "charity" (as Rajib Rajbhandari implied earlier) but about training, capacity building, and empowerment for Nepal. This former employee went on to explain that the culture of the new BPC felt completely different. Gone was the sense of people committed to an institution and working as a team to advance its principles and mission. In the new BPC people were much more likely to mimic their bosses' priorities and work to advance their own narrow interests, even at the expense of colleagues or the company itself. No longer was there a sense of calling to something larger than one's own self-interest. Therefore, rather than behaving ethically as part of a team working for a common goal, individuals turned to self-promotion, like their bosses. The new BPC went from being an "institution" to being a "company" and in the process lost the collective, principled, nation-building mission or ethic that had attracted many dozens of talented and socially minded Nepalis over the previous decades.

No doubt not all BPC employees felt it but eventually this broad wave of emotional malaise reached even the top levels of BPC management. In February 2004, even as Egil Hagen was being pinned under a landslide of criticism, Suman Basnet, the deputy general manager, suddenly announced his resignation. When a position at Winrock International opened up, Basnet grabbed it. Email communications from the time suggest that Basnet's departure was a surprise for everyone. For the IKN side, it was a blow to their hopes of keeping the old BPC culture alive in the new, especially with Egil Hagen being pushed out. For Gyanendra Lal Pradhan, Basnet's exit nixed his plans to have a hopefully more pliable Nepali employee replace Egil Hagen as general manager.

In fact it was precisely this looming clash of cultures that drove Basnet out the door. In an interview with me, Basnet discussed how he had agreed to join the new BPC in hopes that he and others could recreate something of the inspiring, visionary, innovative ethos that the company had developed prior to nationalization. But he had been disappointed. He described being caught between "strong personalities," some of which he admired and others that he found deeply troubling. With his friend Egil Hagen being thrown out of the company, he knew that all of the contradictions would soon rest on his own shoulders—a burden that he refused to carry. Faced with having to deal with Gyanendra Lal Pradhan, Basnet knew that, in his words, "I either had to toe his line, or fight it. And I felt that I could do neither. So I said that it's better that I leave." His decision had to do with more than just "strong personalities." Basnet described how the new BPC was increasingly becoming an ethical minefield with the SEL investor–owners often promoting their own individual business interests over those of BPC (as I will discuss later). When it became clear that Gyanendra Lal Pradhan expected Basnet to serve as a "yes-man" under his direction, Basnet said no. "He was driven less by Odd's philosophy and more by profit motives, making sure that BPC and himself earned as much money as possible. So I just quit."

With his friend Suman Basnet now gone, Egil Hagen too began looking for a way out. In mid-April Hagen wrote a personal letter to Padma Jyoti noting his awareness of, and discomfort with, the criticisms being leveled against him. In this letter Hagen indirectly suggested that he might be willing to step down. Ratna Sansar Shrestha adamantly opposed Hagen's resignation insisting (with good reason) that Hagen was being blamed for conditions that were not entirely his fault. But for Hagen, questions of fault hardly mattered: he was miserable and wanted out. In a May 3 (2004) email to IKN GM Tor Einar Ravnevand, BPC general manager Padma Jyoti wrote to say that the SEL investors had decided to ask Egil Hagen to step down and to establish a new Executive Committee with Gyanendra Lal Pradhan as chair. In reply, Ravnevand wrote: "We are sorry that it has not functioned satisfactorily with

Mr. Egil Hagen as GM and we accept your decision about this." In internal IKN communications Ravnevand expressed strong disapproval of Hagen's departure but admitted that there was nothing IKN could do aside from, perhaps, request that Hagen at least be retained as "technical consultant" to the new BPC general manager. Odd Hoftun wanted to push back against the SEL decision but conceded that it was too late. By early June Hagen had essentially resigned, informing his staff that he had been dismissed and that he was handing over responsibilities to a new acting general manager, R. B. Karki. Announcing that he was "completely exhausted and needs to get away," Hagen and his family left Nepal by the middle of June 2004.

That summer another delegation of SEL members (and their wives) headed for Oslo for meetings with the IKN board to try to smooth out their crumpled relationship. But for IKN it was clear that, starting from a weak position, they had been outmaneuvered and left with no choice but to concede defeat. Hoftun's hopes of maintaining some level of influence on the new BPC through the appointment and supervision of top management figures were finished. With grudging support from the IKN board (and against the wishes of Ratna Sansar Shrestha) by late 2004 the BPC board approved a new Executive Committee—aimed at addressing what it said was poor company performance since privatization. With this move power shifted into the hands of the new Executive Committee chairman, Gyanendra Lal Pradhan. Pradhan was to report to the BPC board but now had explicit supervisory authority over the entire BPC management structure, including the general manager. But, as the following sections will make clear, already by the time Pradhan took over managerial control controversial developments were under way that eventually ended in his acrimonious departure from BPC.

REPAIRING JHIMRUK

On March 31, 2002, Maoists had bombed the Butwal Power Company's (BPC's) Jhimruk power plant. The already ill-fated and locally unpopular facility was left severely damaged, idle, and inoperable. Although there is never a good time for a bombing, this one had come at an especially unfortunate moment as the Interkraft Nepal (IKN) and Shangri-La Energy

Limited (SEL) investors, after years of frustration, were finally moving toward the successful conclusion of their bid to purchase BPC from the Nepal government. The IKN–SEL bid had been accepted but, with the destruction of Jhimruk, BPC's assets were suddenly diminished, throwing the entire bid and privatization process into confusion. Months later the Norwegian government broke the stalemate that had ensued over who would pay for Jhimruk's repairs. Norway offered to provide around USD 3 million for Jhimruk's rehabilitation, but only on the condition that BPC be privatized. The grant would only be issued to an *independent* BPC. As described in the previous chapter, it was partially the threat of losing this money that helped force the Nepal government to finally approve BPC's privatization. Thus, in early 2003, the first order of business for the new BPC was to mobilize the Norwegian grant money to get Jhimruk repaired, back on line, and generating money.

Unfortunately, repair efforts quickly devolved into a bitter struggle between competing visions, values, and priorities from BPC's leading parties, IKN and SEL. In the conflict between IKN and SEL over the Jhimruk Rehabilitation Project (JRP), as it was officially known, both sides had plenty to be unhappy about. While the SEL investors accused Hoftun of irresponsibly overstepping his authority by negotiating on BPC's behalf without board approval, the IKN side accused SEL of compromising on quality in the name of cost reduction and, more seriously, of engaging in unethical business practices by steering BPC business to their own private companies.

Soon after Jhimruk was bombed in 2002, Hoftun had begun working on how to bring about the plant's restoration as quickly as possible. Hoftun felt responsible to get the ball rolling even before privatization was completed: waiting to initiate the repair process until *after* the handover would only add to the delay, and the longer Jhimruk remained idle, the less profitable BPC would be. For Hoftun it was a foregone conclusion that BPC's daughter company, Nepal Hydro and Electric (NHE), would take the lead in repairing Jhimruk, using new electromechanical equipment from NHE's long-time Norwegian collaborator (and partial owner) Alstom. Alstom had been an important contributor to technology transfer to NHE and was, in fact, even then in the process of shipping an entire decommissioned Norwegian generator repair workshop to Nepal, thereby significantly enhancing NHE's functional capacity in Nepal's hydropower services market. As general manager of the IKN board, already in mid-2002 Hoftun had approached Alstom, Jhimruk's original equipment suppliers, for estimates on new machines and control panels. The Norwegian government had based its grant for Jhimruk repairs on these estimates.

In an email to Ratna Sansar Shrestha, Hoftun acknowledged that, once the grant was announced, he had requested that Alstom start working on the project even without a "firm contract. I said that if in the end no contract was made, then JRP [the Jhimruk project being run by BPC] would pay for the work done." According to Hoftun, when he informed others—including Gyanendra Lal Pradhan—of this "gentleman's agreement"[4] with Alstom, he got no response. He took this as approval. Although surely well intended and coming from a person with unsurpassed understanding of BPC's past and present needs, Hoftun's unilateral decision to engage Alstom's services without a contract or BPC board approval seems unwise and perhaps even unethical, especially given the already tense relations between the IKN and SEL investors. No doubt Hoftun assumed that the new BPC would operate according to the same logic and priorities of the old BPC.

But the SEL investors, now the majority owners of BPC, had different ideas. Deep in debt and faced with imminent loan payments, the SEL partners looked at the Norwegian lump-sum Jhimruk grant as an opportunity to practice frugality with the aim of finishing up the repair process with a cash surplus that could be used to help pay down their debt. The money was theirs. Why spend it all on Jhimruk repairs if they did not have to? Furthermore, as newcomers to BPC and the world of hydropower, the SEL owners—unlike Hoftun—felt no particular sense of allegiance to, or partnership with, BPC's past suppliers. Very quickly the two sides were working at cross purposes with hurt feelings and bruised egos all around.

By March 2003 Gyanendra Lal Pradhan and the other SEL investors were aware that (at Hoftun's behest) Alstom had already begun preparations for the Jhimruk repair process but they decided to approach other electromechanical

---

[4] A phrase Hoftun later used to describe his agreement with Alstom in a document entitled "Updated Brief History of the Melamchi Diversion Project," dated December 2007.

suppliers to see if they could spur competition and lower costs. Gyanendra Lal Pradhan approached Wassercraft Volk (WKV), the German hydropower supplier for whom his nephew (and Hoftun's long-time collaborator) Balaram Pradhan now worked as marketing agent for Nepal. At the same time SEL representatives approached Indian and Chinese equipment suppliers who offered to sell the necessary machinery for half or even a third of what the European companies charged.

With a variety of much lower bids in hand, Gyanendra Lal Pradhan and Rajib Rajbhandari resumed talks with Alstom looking to negotiate a lower cost, but the results were ugly. Exactly what happened is unclear but apparently Alstom was in no mood to compromise. Based on their earlier bid, and Hoftun's verbal go-ahead, Alstom engineers had already begun design work on the Jhimruk repair project. No doubt they were unhappy to have BPC officials show up asking to revise what Alstom thought were agreed-upon conditions—especially when being asked to "compete" with Indian and Chinese manufacturers whose products were simply inferior by any objective standard. The Nepali negotiators found Alstom to be brusque, uncompromising, and (intentionally or otherwise) insulting. For their part, Alstom was sufficiently upset over what to them amounted to a breach of contract (though there was no official contract) that they issued a public complaint to Norad, the Norwegian funders of the Jhimruk repair project. Having been briefed on the meeting between the BPC representatives and Alstom, BPC Board Chairman Padma Jyoti wrote to Hoftun:

> From what we heard I could not understand how a company [Alstom] trying to sell its products could act like this with a buyer and hope to get orders. Anyway, let us put it down as cultural differences and miscommunication. We must repair our relationship with Alstom and I think they should also try to be more customer-oriented.

Jyoti argued that "competitive bidding" is "always better" and criticized Alstom for taking the "emotional step" of complaining to Norad, thereby "leading to bad publicity."

What Padma Jyoti and the other SEL investors understood to be a relationship between "a company" and "a buyer" was for Odd Hoftun a relationship between two long-time collaborators each with a significant sense of obligation to the other based on decades of personal contact and partnership. Admitting that, if his verbal agreement with Alstom fell through, "I will suffer a big personal liability," Hoftun watched with growing alarm and anger as developments unfolded. On March 19 Hoftun warned that "it would now be very short sighted to desert Alstom as a partner." A few days later he was "flabbergasted" to hear that the SEL group was leaning toward WKV as the equipment supplier and even considering replacing Alstom with Wasserkraft as NHE's official collaborator. For their part, the BPC board approached the Norwegian embassy to ask whether the Norad grant *required* that BPC use a Norwegian supplier and were assured that it did not.

By late March 2003 Hoftun was seething. He accused the SEL-dominated BPC board of "doing their level best to MAXIMIZE [*sic*] their profit from the destruction of Jhimruk" even if it means "hurting BPC in the long run." He charged the SEL investors with being blinded by greed while ignoring "all other extraneous factors like honor, dignity, ethics, etc." Hoftun claimed not to be opposed to competitive bidding but argued that BPC's commitment to Alstom should have been "simply a question of orderly procedures and decent behavior."[5]

Of course Hoftun had good reason to be upset with the BPC board's turn to WKV. WKV had no real experience in Nepal and they were not known as manufacturers of the kind of electrical control systems that were needed (among other things) at Jhimruk. Hoftun argued that whereas Alstom had custom designed the original Jhimruk control panels to exactly accommodate the project's electromechanical equipment and generating conditions, and could, therefore, quickly reproduce the damaged goods with exact copies, WKV would have to "start from scratch offering to design and manufacture

---

[5] Reflecting on the episode several years later, Hoftun was no less incensed. He accused the SEL owners of "boundless greed" that had forced him and others to lose "every illusion we might have about Nepali business ethics." The lesson to be learned, wrote Hoftun, was: "One just has to nail down every detail in legal documents, and not leave anything to gentleman's agreement. There are no gentlemen in Nepal when it comes to pay out money." (From a document entitled "Updated Brief History of the Melamchi Diversion Project," dated December 2007.)

a control system in little more than two months, without at this point having even worked out the technical details with the customer." While Alstom was prepared to hit the ground running, it would take Wasserkraft months to even work through the technical specifications thereby only lengthening the time that the Jhimruk power plant would remain idle and BPC would continue to lose money.

Sadly the whole affair also soured Hoftun's relationship with his old friend Balaram Pradhan, whom Hoftun now suspected of being behind WKV's emergence as the BPC board's favored bidder. Furious, Hoftun wrote to Ratna Sansar Shrestha, "No wonder that Balaram wanted to disassociate himself from IKN—but I have no proof that he was involved in this." To Hoftun, dealings between relatives Gyanendra Lal and Balaram Pradhan were "foul smelling business." Judging by Balaram Pradhan's own correspondence, he was unaware of being a target of Hoftun's frustration (and was unlikely to have been the instigator of BPC's turn toward Wasserkraft). As mentioned in the previous chapter, it was at about exactly this time—with the whole exhausting privatization process having finally been recently completed— that Pradhan wrote to Hoftun affectionately reminiscing about their long partnership and asking for a "souvenir" letter of recommendation to mark the end of this long and trying chapter in their relationship. It took Hoftun more than a year to even acknowledge Pradhan's warm letter.

By the end of March 2003 Hoftun wrote to colleagues in Nepal of his "wish to withdraw as a volunteer advisor to JRP" complaining that his advice was being completely ignored. "I cannot but take this behavior from JRP's side as an insult and a clear sign that my input is not wanted any longer, and that JRP wants to do business in a style I am not used to and happy to be part of." At this, Board Chairman Padma Jyoti asked that everyone reset and refocus on the future "with a positive mind." He also apologized to Hoftun, requesting that he "continue to be our guide from Norway." To Ratna Sansar Shrestha Hoftun reflected—with understatement—on how "I thought things would become easier after we [the IKN group] had taken over BPC and gotten our new management in place, but that has not happened."

In the coming months the relationship between the IKN and SEL investors only grew more tense as it became clear that the SEL side's commitment to "competitive bidding" was not as deep as they had earlier claimed. Specifically, when dealing with outside contractors, competitive bidding was, in Padma Jyoti's words, "always better." But when dealing with supply contracts that could be filled from within companies that made up the SEL consortium, the BPC board felt no compulsion to seek outside competitive bids. The SEL investors represented a wide array of Nepali companies offering goods and services that BPC could, and did, require. It was soon clear that BPC board members were perfectly comfortable with granting non-competitive contracts to companies within the SEL "family."[6]

The matter came to a head already in May 2003 when the IKN representative to the BPC board, Ratna Sansar Shrestha, accused Gyanendra Lal Pradhan of violating Nepal's Company Act. At a board meeting, Shrestha learned that the contract signed with WKV for Jhimruk repairs did not include control cables, which were now necessary to acquire, and quickly. Suspiciously, the board quickly authorized the purchase of cables from Trishakti Cable, a Nepali company owned by Gyanendra Lal Pradhan. When challenged at the meeting, Pradhan defended the decision arguing that BPC contracting with Trishakti was no different from BPC contracting with NHE. But Shrestha countered that, whereas NHE was an official subsidiary of BPC (with BPC its majority shareholder), Trishakti Cable had no legal subsidiary status. According to Nepal's Company Act, awarding contracts to any company owned by a sitting board member was forbidden by law. Hoftun affirmed Shrestha's efforts to "nip the evil in the bud" but it was soon clear that the desire to steer business to their own companies was already deep rooted within the SEL collective. Defending the board's actions, Chairman

---

[6] Here it is telling to compare the new BPC's dealings with companies within the SEL "family," and those with NHE, the Nepali electromechanical equipment manufacturer founded by Hoftun in 1985. In purchasing BPC from the Nepal government, the SEL investors had also acquired a majority share in NHE. Yet while the new BPC board authorized all the business they could with companies owned by SEL investors, BPC's dealings with its own subsidiary, NHE, dwindled to almost nothing. Before BPC's privatization more than 75 percent of NHE's business came from its parent company. After privatization, NHE's business with BPC fell to less than 5 percent. Rather than promoting its own daughter company, BPC patronized mainly Chinese equipment suppliers. Presumably the benefits of dealing with Chinese companies were greater than bolstering the competitiveness and profitability of NHE.

Padma Jyoti acknowledged Shrestha's concerns but dismissed them saying that BPC board members should "not be in the business of scoring legal points or winning debates on ethics." No one wants to be on the wrong side of the law, he added, but "our first loyalty has to be toward our own group and the job at hand."[7] Soon Padma Jyoti would be forced to acknowledge that some of the SEL board members were motived less by "loyalty toward our own group" than by loyalty to their own private interests.

Through the summer and fall of 2003 Jhimruk rehabilitation proceeded under contract with WKV but the tension between cost reduction and prudent investment continued. Hoftun warned repeatedly that cutting corners on the repair project would only be detrimental to operations in the long run. And it was not only Hoftun who was concerned. Email exchanges between BPC Nepali project engineers and Hoftun describe how they were constantly under pressure from management to lower costs and cut corners. Frustrated, one engineer summed up the situation as follows: "For BPC and its owners the key thing now is cash flow. They are interested to invest in improvements that will improve cash flow." Or, as another observer put it, BPC management aimed to "strip the company of its liquid assets" which, in this case, meant spending as little of the Norwegian JRP grant as possible in order to service debts. Rather than diverting profits from their other companies to pay off their BPC debt, the SEL investors were willing to cannibalize BPC even if it left the company seriously weakened and unable to accomplish the development goals laid out in the company's business plan.

In the meantime, Alstom—acting on Hoftun's earlier promise to pay for Alstom's services even without a formal contract—billed BPC for NOK 500,000 for design work it had carried out prior to BPC's shift to Wasserkraft. They also accused BPC of using design elements that Alstom had included in a "confidential contract proposal" that was never finalized. Hoftun took the

---

[7] Jyoti's admission of "loyalty to our own group" was also in direct violation of the sales and purchase agreement (SPA) that the SEL investors had agreed upon as part of the legal basis for their purchase of BPC. Clause 9.2 of the Covenants section of the SPA promises to pursue competitive bidding for all BPC contracts and specifically stipulates that BPC "shall not grant favorable or concession terms to any of the shareholders, their associates, affiliates, connected persons or private interests."

BPC board's refusal to pay as a further personal humiliation. By the time the Jhimruk plant finally went back online in late 2003, industry insiders guess that the SEL group had spent as little as half of the original Norwegian JRP grant.

"UP-GRADING" ANDHI KHOLA

If the Jhimruk Rehabilitation Project (JRP) was one prime example of conflicting visions for the Butwal Power Company (BPC) between the minority Norwegian Interkraft Nepal (IKN) investors (and their Nepali colleagues) and the majority Nepali Shangri-La Energy Limited (SEL) investors, another illuminating flashpoint was the "up-grading" of BPC's Andhi Khola hydel plant. Here again a Hoftun-inspired development and business philosophy clashed with the values and priorities of the SEL investors. And here again those looking to maintain Hoftun's ideals lost—even though, in the long run, their recommendations might have charted the best course.

In the summer of 2003 BPC board member and SEL group leader Gyanendra Lal Pradhan formally proposed that BPC could significantly increase its Andhi Khola hydel plant's output by installing new, larger turbines and generators. New equipment would mean a total of 9.4 megawatts (MW) peak output as opposed to the plant's current 5.1 MW. Andhi Khola's electromechanical equipment—refurbished Norwegian machines dating from the 1920s—were inefficient, unreliable, and increasingly difficult to get parts for, Pradhan argued. What's more, water was available that was not being adequately used: bigger turbines would more effectively use the facility's actual water resources. According to Pradhan, "upgrading to a higher capacity by the installation of new electro-mechanical equipment and controls is probably the only solution" (Bhandari and Pradhan 2006: 80).

Presented at a BPC board meeting in 2003, it was this proposal by Pradhan that elicited BPC General Manager Egil Hagen's blunt retort: "That is sheer nonsense!" Hagen's regrettably undiplomatic reply (in an already combative atmosphere) had the unwanted effect of overshadowing Hagen's main concern with what he believed were faulty mathematics in Pradhan's engineering proposal.[8] Yet, as several observers noted, the *tone* of the debate quickly overrode its *rational content* with each side digging in for a fight over who was stronger. No longer a matter of rational debate, it became "an issue of pride between the two sides," as one person recalled.

At question was *not* whether BPC could get more power out of Andhi Khola, but whether replacing the plant's major components was a cost-effective way of doing so. Would the benefits of a large-scale upgrade outweigh the costs? Hagen, Hoftun, Ratna Sansar Shrestha, and others on the IKN side argued that installing new equipment at Andhi Khola was a bad idea. The cost of new equipment and the labor to install it, plus the loss of revenue when the plant was shut down for upgrading, would be a big drain on BPC's limited resources. But most importantly, they argued, upgrading Andhi Khola would really only increase the plant's wet-season production: bigger turbines could better handle monsoon flow rates but would have very little impact on the plant's ability to generate power during the dry season when cursed "load shedding" was the only solution to the national power deficit. Rather than invest their limited money on an upgrade that would only generate more power at a time of year when there was already a glut, they argued that it would be much better to invest in *new* hydel projects that would increase BPC's dry season production—when rates and demand were highest.

Hoftun readily acknowledged that Andhi Khola suffered from various inefficiencies but argued that, with routine maintenance and perhaps some minor upgrades, it made much more sense to let the plant continue to generate reliable profits to be invested elsewhere than to sink major funds into renovations that would be of limited value. Hoftun admitted that Andhi Khola's 5.1 MW output was mainly a consequence of the available used, essentially free, Norwegian equipment that Nepal Hydro and Electric (NHE) had initially refurbished and installed at Andhi Khola. The site's actual capacity was more like 9 MW. Already in 2001 Hoftun had recognized that the Andhi Khola equipment needed periodic maintenance and even replacement of major parts. For example, Hoftun noted that the plant's Pelton turbine runners needed to be replaced but argued that the

---

[8] Specifically regarding how much increased power production the equipment upgrade would accomplish.

needed buckets could be cast in India and milled by NHE for a fraction of the cost of new equipment. Similarly, he noted that at least two of the plant's three generators needed to be rewound but, again, argued that this was work that could be done at relatively low cost by NHE once they had installed the generator repair workshop (from a decommissioned generator manufacturing plant in Norway) that Alstom was in the process of donating to NHE. Hoftun estimated that addressing these two issues could raise Andhi Khola's output by 10 percent. A few years later Hoftun worked up detailed plans showing how further relatively minor investments in new bindings and other components for the generators, plus new, more efficient turbine runners could boost the plant's productivity another 10–15 percent. If BPC's new owners wanted more out of Andhi Khola, here were some ways to do it in a cost-effective manner.

But Hoftun's advice fell on deaf ears. His repeated proposals and repeated visits to Nepal to meet with board members seem to have only hardened opinions. By the summer of 2006 one of the SEL board directors reportedly complained in public that Hoftun's "level of interest is unhealthy." Couched as concern for Hoftun's health, the implication was clearly that his input was unwanted. A few months later, when even Hoftun's comrade Ratna Sansar Shrestha encouraged him to back off, Hoftun replied pleadingly, "I promise that this shall be the last time I get involved in BPC affairs. But this last time I do dearly wish to see my proposal accepted by those in authority—because this is good and profitable for BPC and for Nepal."

Ironically, it was in the midst of this increasingly bitter dispute over Andhi Khola's future that the UNESCO International Hydropower Association presented the Andhi Khola project with its Blue Planet Prize recognizing the original project for "excellence in socio-economic benefits and capacity building." The award praised the project for its creative use of appropriate technology to make power affordable for consumers, its integrated development approach, and its emphasis on training Nepali manpower. In other words, the award praised the *old BPC's* development ethic even as the *new BPC* was heading in considerably less socially oriented directions.

As BPC's executive director, Gyanendra Lal Pradhan went to Montreal to accept the award but Odd Hoftun, the specially invited guest of honor, stayed at home. Later he explained to me:

I knew that Gyanendra stands for the opposite of those policies that the Andhi Khola Project has demonstrated, and which did qualify the project to receive this award.... It would have been simply too embarrassing for me to be present there, keeping in mind the serious and well-known disagreements that he and I had concerning the future direction of BPC. So I refused to attend.

However, in March 2006 Hoftun did accept an invitation to speak at the African Ministerial Conference on Hydropower and Sustainable Development in Johannesburg in recognition of the Blue Planet Prize. There Hoftun delivered an impassioned defense of his bottom-up, cooperative, capacity-building approach to hydropower development. Rather than following "directives from above," or pushed by profit motives "where the time perspective is too short," Hoftun advised his listeners to have "the faith and courage to go and plant a seed."[9] From tiny embryos, successful development has to take root in local soil, a process that cannot be hurried or dictated. Likely few of his listeners knew that the values Hoftun extolled were even then being seriously undermined.

Meanwhile, as Gyanendra Lal Pradhan and the rest of the BPC leadership continued to push ahead toward a major overhaul at Andhi Khola, Hoftun and many others began to wonder about the purity of the motives behind the plan. If the cost–benefit analysis was not in favor of a major upgrade in terms of *BPC's* interests, were there other interests at play for whom the costs *would be* a benefit? Would Pradhan and perhaps other SEL owners benefit from the project even if BPC itself did not? All of the independent observers I spoke with speculated that, in the end, profits to other parties from within the SEL collective—and especially Gyanendra Lal Pradhan—must have played a major role in how the events unfolded.

Part of this controversy centered on what kind of electromechanical equipment would be purchased for the Andhi Khola upgrade—German or Chinese? The German equipment was more expensive but more compact. Installing it would require no alterations to the existing powerhouse or

---

[9] From a presentation to the African Ministerial Conference on Hydropower and Sustainable Development, Johannesburg, South Africa, March 2006.

tunnels. By contrast, the Chinese equipment was cheaper but considerably larger than the new German or old Norwegian machines. Installing the new Chinese turbine-generator sets would require actually re-excavating and enlarging Andhi Khola's underground powerhouse cavern and doing the same to the tailrace tunnel. Therefore, simply to *install* the new Chinese equipment would mean major, costly, time-consuming structural renovations—in fractured, unstable rock that had been difficult to deal with the first time and would be again. While the machines themselves would be cheaper, installing the Chinese equipment would be much more expensive overall. In spite of these drawbacks, Gyanendra Lal Pradhan chose the Chinese equipment—a move that many suspected of involving foul play. The BPC board approved the decision but later came to regret it.

One hydropower insider compared Pradhan's move to an analogous case regarding the Nepal Electricity Authority's (NEA's) renovation of its Trisuli power plant, built in the 1960s using European and Japanese electromechanical equipment. When NEA received money from the World Bank to upgrade Trisuli, NEA officials decided—over the objections of their own engineers—to scrap the original equipment and replace it with inferior Chinese machines. The engineers argued that the money could have been used much more effectively in upgrading the project's civil works or, better yet, investing in new construction. In the end, the Trisuli "up-grade" did virtually nothing to increase the plant's dry-season production and the World Bank's money was basically squandered. The very matter-of-fact assessment that many people told me was that NEA officials had pocketed substantial kickbacks from the Chinese equipment suppliers and that this was more than enough reason to go the Chinese route.

The same person who told me about the Trisuli case noted that the Andhi Khola upgrade was also funded by the World Bank (International Finance Corporation [IFC]) and laid considerable blame for both projects at the World Bank's feet. He argued that, in both cases, the World Bank should have much more carefully examined the basic utility of the proposed upgrade—especially against the possibility of using the same funds to invest in new construction that would actually enhance Nepal's dry-season power output. Even more pointedly, he argued that the World Bank should have questioned key decisions made by Trisuli and Andhi Khola leaders. Instead, in both cases the Bank turned a blind eye to shady dealings. Even worse, in the case of BPC's Andhi Khola upgrade, as cost estimates for the project rose substantially over the years (as the price of powerhouse excavation and other major changes to the civil works were added to the mix), instead of pulling the plug on their financing, the World Bank increased its loans while also demanding new security measures that only made financing even more costly.

According to BPC's initial cost estimates, the Andhi Khola upgrade should have cost NPR 800 million (about USD 8 million) or around USD 2 million per added megawatt of production. By contrast, Odd Hoftun had managed to build *the entire* original Andhi Khola project (head works, tunnels, penstock, and so on) for less than USD 1 million per megawatt. Here just a simple upgrade would cost twice as much in terms of increased output. But things only got more expensive from there. By the time renovations actually began at Andhi Khola, the cost estimates had been revised up to USD 12 million (now USD 3 million per added megawatt). When the upgraded Andhi Khola finally went back into service, the total cost had risen to USD 14 million (USD 3.5 million per added megawatt).[10] Furthermore, the renovations that had initially been projected to take six months to complete had taken two years: two years during which one of BPC's most reliable and profitable assets had been shut down—adding insult to injury.

Observers I spoke with speculated that another reason for why BPC's leadership decided to push ahead with the seemingly nonsensical Andhi Khola upgrade was because members of the SEL investor collective stood to profit from construction contracts with their own private companies.[11] Contracts for construction products such as concrete and steel rebar went to Nepali firms owned by SEL investors. By this logic, short-term profits made by companies within the SEL family justified long-term damage to BPC itself. Costly international financing seemed justified if some of the money ended up in the pockets of SEL investors.

---

[10] This final cost per added megawatt is almost twice the average cost of Nepali private sector hydropower construction (Karki 2017: 121).

[11] Shrestha (2017: 141) notes that this is a common phenomenon across the hydropower development industry in Nepal as contractors make extra money off of construction at the expense of funding agencies.

At the end of the day, BPC had spent USD 14 million (considerably more than the SEL–IKN investors had spent to buy the *entire company* from the government) and lost two years of Andhi Khola revenue, all to increase its total corporate power production capacity by only 4 (phenomenally expensive) MW. Rather than invest its limited cash resources and available loan monies in new construction, BPC had transformed a debt-free, profit-producing asset (the 5.1 MW Andhi Khola plant) into a bloated white elephant producing power that is now so expensive (because of heavy debt repayment obligations) that it will take years for it to return to profitability. Perhaps not surprisingly, by the time Andhi Khola was finally back on line (in early 2015), Gyanendra Lal Pradhan had long since parted ways with BPC.

KHUDI PROBLEMS

Occurring simultaneously with the controversial Jhimruk Rehabilitation Project (JRP) and early stages of the Andhi Khola "up-grade" between 2003 and 2006, what finally alienated Gyanendra Lal Pradhan from even his fellow Shangri-La Energy Limited (SEL) investors were the shady circumstances surrounding the development of the Khudi hydropower project. In the late 1990s a Canadian engineering firm had teamed up with a Lamjung-based developer to design a 4 MW project on a tributary of the Marsyangdi River near the famous Annapurna Circuit trekking trail.[12] Due to political turmoil and trouble with financing, the project languished until 2003 when Gyanendra Lal Pradhan, representing the newly privatized Butwal Power Company (BPC), decided to invest—over the objections of Odd Hoftun.[13]

---

[12] "Khudi Hydropower Project," Canadian Consulting Engineer, October 1, 2008, https://www.canadianconsultingengineer.com/features/khudi-hydropower-project/ (accessed August 17, 2017).

[13] Surveying the Khudi project's design plans and location, Hoftun argued that the site was less than ideal. Because the rock was unsuitable for tunneling, the project would require relatively complex civil works, including 2.5 kilometers of exposed penstock and a complicated inverted siphon design, that would make the power plant both relatively costly (for the small amount of power it would produce) and susceptible to flooding. All of these premonitions proved to be correct.

In the new Khudi Hydropower Limited (KHL), BPC held a 60 percent (majority) investment share and Pradhan took charge of the KHL board—with no significant oversight from the BPC board. Himal Hydro built the project—including 2.5 kilometers of exposed steel penstock—between February 2005 and the plant's commissioning in December 2006. Completed on time and within budget, Khudi won its Canadian design partner an award in 2008 for the project's local "empowerment" impact.[14]

But even before Khudi's commissioning, trouble was brewing within BPC's management structure. By early 2006 signs of discontent were beginning to surface from within the SEL investor group regarding Gyanendra Lal Pradhan's leadership. Up to that point his SEL colleagues had been happy to leave management and technical responsibilities to Pradhan—the only one among them with engineering credentials. Yet over time the SEL investors grew wary of Pradhan's private dealings on the sidelines, and even within BPC. It was one thing to channel BPC business contracts to one's own private company—as in the Trishakti affair (in which BPC purchased electrical cables from one of Pradhan's companies for JRP). But it was another thing to use one's administrative power to pursue personal enrichment at the expense of the company, which (based on their subsequent actions) many within SEL now suspected of Pradhan.

From his position on the BPC board, and via various reports from BPC insiders, Ratna Sansar Shrestha watched as Gyanendra Lal Pradhan's authority eroded. In May 2006 Shrestha reported to Hoftun that Pradhan "has already lost a substantial amount of credibility amongst his SEL colleagues." A few weeks later Shrestha reported rumors circulating to the effect that Pradhan had received hefty "commissions" from the Chinese company that he had chosen for the Khudi electromechanical equipment contract. A year later, by April 2007, SEL members were in open revolt over Pradhan's leadership. Two of the three BPC representatives on the KHL board (the third being Pradhan himself) resigned in protest, a move that Shrestha understood to confirm reports of Pradhan's financial improprieties. Shrestha reported news of a campaign to "persuade" Pradhan to resign from his KHL and BPC board positions and on May 17, 2007, he reported

---

[14] "Khudi Hydropower Project," Canadian Consulting Engineer.

to Hoftun the still unofficial news that "GLP has been forced out from the position of MD of SEL." Pradhan's Khudi project had been the last straw and, as if to add insult to injury, in its first several years of operation the Khudi civil works twice suffered significant damage due to flooding (and poor design) considerably adding to the project's overall cost and debt. Ten years after its commissioning the Khudi project had yet to earn BPC any dividends. A combination of high cost and low output meant that Khudi was generating electricity but not generating profit.

In some ways Gyanendra Lal Pradhan managed to get the last laugh, literally at BPC's expense. Once he had been stripped of his power within the corporation, Pradhan's SEL colleagues and co-investors could have either allowed him to remain a marginalized shareholder, or buy out his shares and be rid of him altogether. In a move that some observers found to be unwise, they chose the latter. The SEL collective bought Pradhan's shares at a rate vastly in excess of Pradhan's original investment leaving him with a huge windfall profit. "That guy made a killing!" one industry insider told me. Pradhan was handsomely rewarded for making a series of decisions that weakened BPC and enriched himself. As if to confirm his skills at self-promotion, in 2010 the journal *Hydro Nepal* awarded Gyanendra Lal Pradhan its Hydro Nepal Excellence Award for "outstanding and untiring, tenacious and entrepreneurial efforts in hydropower and energy development in Nepal." The award cited Pradhan's leadership of BPC, under which "performance increased significantly," and the Blue Planet Prize for Andhi Khola. That same year Pradhan founded Hydro Solutions Limited, a private hydropower development company which, along with many other recent entries into the field, was in direct competition with BPC.

THE NEW KATHMANDU HEADQUARTERS

In 2006, under Gyanendra Lal Pradhan's leadership, the Butwal Power Company (BPC) undertook another controversial project: purchasing land and erecting a costly new administrative headquarters building in Kathmandu. BPC had long rented office space in the capital while its allied companies maintained offices in both Kathmandu and Butwal.

Consolidating offices in one place made sense in terms of hiring and retaining Nepali professional staff. But the decision to make a costly move came at a difficult time for the company when it was already burdened with deep debt (from the privatization process, the Khudi project, and the Andhi Khola "up-grade"). With only seven years to pay off its privatization debt, BPC had to pay out annual dividends of between 20 and 25 percent just so that the Shangri-La Energy Limited (SEL) investor–shareholders could meet their debt obligations. (In theory SEL investors could have used profits from their *other* corporate enterprises to help pay off BPC debt. But with the Nepal economy in shambles during and after the People's War, the SEL investors could not, or chose not to, risk using outside profits to pay for their already risky venture into hydropower.) Asking BPC to pay for itself in short order meant that there was little or no dividend money available for reinvestment in the firm. Without dividends, the few moves that the company made (Khudi, the Andhi Khola "up-grade") were based on loans that put the company in even deeper debt and, as discussed earlier, saddled with unproductive assets.[15]

Nevertheless, BPC went ahead with a costly move that, many argued, was unnecessary and would only sink the company deeper into debt. Advocates argued that BPC would benefit from centralizing its administration, along with its corporate spin-offs (who would be required to rent space from BPC[16]), into one building—rather than paying rent in dispersed locations. Plus, with land prices rising fast, Kathmandu Valley real estate was a good investment. Why not use a new building to imprint a strong corporate image on the Kathmandu skyline? Critics pointed to the fact that there were already plenty of rental spaces in the valley[17] and that new construction (with its huge losses due to depreciation and debt payment) was hardly the best choice for a cash strapped company. BPC tore down a former United Mission to Nepal

---

[15] Ironically, the hugely expensive "up-grading" of Andhi Khola transformed that facility from a profit-generating to a profit-consuming "asset."

[16] In fact some of these companies, such as Nepal Hydro and Electric (NHE), refused to occupy the new building because of the high rents that BPC demanded for office space.

[17] The 2015 earthquakes actually turned out to be good for BPC given that much of the available office space in the valley was damaged, which meant that BPC's new construction was suddenly more valuable.

(UMN) high school building in Buddhanagar (next to the foul-smelling Dhobi Khola) and built a ten-story glass tower with 3,400 square meters of floor space, including a posh auditorium on the top floor that is used once a year for BPC's annual shareholder meetings.

The BPC management's decision to add a costly office tower to the company's list of financial burdens again raises uncomfortable questions about the SEL investor–owners' priorities. Numerous industry insiders I interviewed pointed to a conflict of interest for BPC's SEL shareholders over whether to prioritize BPC's corporate interests, or to treat BPC as a client for products and services purchased from their other individually held companies. When the BPC board chose not to pursue competitive bids for corporate services available from within the SEL collective for the Jhimruk Rehabilitation Project; or when the BPC board approved a vastly more expensive version of the Andhi Khola "up-grade" project (requiring major powerhouse reconstruction) that would award supply contracts to SEL investors; or the decision to build a palatial office building that would, again, contract with SEL investor firms—the question becomes, "Why?" Whose interests are being served by decisions that apparently hurt BPC's bottom line in favor of profits for other companies within the "SEL family"? A charitable view would see these moves as mutually beneficial for all concerned. But a cynical reading sees SEL investors mining BPC for wealth to serve their own private interests.

Somewhere between these readings is a more resigned acknowledgment of managerial steps (and missteps) taken in good faith in the midst of extremely challenging economic and cultural contexts. Even vocal critics of BPC management decisions admit that running a business in Nepal— and perhaps especially a complicated hydropower development business— is hugely challenging. "In Nepal a manager has to have more than just managerial skills. They have to be able to deal with the government, with law and order problems, with strong trade unions—you have to deal with so much more than simple management!" explained one person. Another (a former BPC executive) laughed and said, "Even the fanciest MBA equips you with only part of the skills needed." He went on to explain that the whole environment is just too chaotic and unstable to allow for anything approaching "normal," much less "optimal," business practices.

Several observers noted tensions between managers and engineers at BPC. Especially after the departure of Gyanendra Lal Pradhan (and his engineering credentials), and through a succession of ineffective CEOs brought from the outside, BPC management tended to rely more and more on in-house engineers. The engineering function is obviously vital to a hydropower development company. But that function is completely different from, and sometimes at odds with, the managerial function. An engineer's job is to solve problems. But someone else has to decide which problems are worth solving.[18] Feasibility has little to do with financial justifiability or, as one person explained, "Quite simply, the benefits should be larger than the costs" if a business is going to be sustainable. This person pointed to the Andhi Khola "up-grade" and the new corporate headquarters building as instances of decisions driven by engineering logic, not business logic. In this view BPC has sometimes failed to integrate sober cost–benefit analysis into its management decisions. "If that's missing, then you're in a mess. If you listen too much to the engineers, then you're in trouble."

BPC also struggles with basic bureaucratic efficiency. In 2013 BPC hired a Swiss business consulting firm to evaluate the company's management and human resources performance and policies. The resulting "HR Process Review" document was almost unbelievably scathing. In a four-year period ending in July, the analysis found that BPC's "total operating income grew by 35%, while total expenditures grew by 72%. At the same time generation expenses grew by 98% and administrative expenses grew by 41%."[19] With around 300 employees, including 60 at the Kathmandu headquarters, the report found that BPC was "highly overstaffed" and inefficient. At headquarters, "one gets the impression of an astounding low level of visual professional activity amongst the staff…. Group discussion, newspaper

---

[18] Shrestha (2017: 132–133) refers to this cost management function as "financial engineering" and credits Odd Hoftun with introducing the idea (if not the term) into Nepal's hydropower sector.

[19] This quotation and those following are from a management report prepared by Swisscontact Senior Expert Corps entitled "HR Process Review: Findings and Recommendations," prepared by Walter Zahnd, dated December 9, 2013.

reading, and obvious non-job related activities were observed throughout." The report continues:

> <u>Communication is an issue</u> – between organizations and between layers of management and employees. Even within workgroups at SBU level, it became apparent that employees have little knowledge about the activities of their peers.
>
> <u>Important management skills are missing</u>: people management, decision making, performance management, planning and organization, project management, change management.
>
> Managers seem to be aware of issues but seem to be "waiting for instructions" from above. No personal initiative is taken and there is no sense of urgency. "Assuming responsibility" is not a strength of BPC managers.
>
> <u>Employees and managers do not seem to strive for performance or for improving current performance</u>. Most employees seem to accept the status quo and see no need for any changes.
>
> Interviews with employees and managers showed a high degree of <u>employee dissatisfaction</u>. In fact, several managers indicated that they are actively searching alternative employment. While exits are welcome in an overstaffed organization, the risk exists that the most competent and knowledgeable staff will eventually leave. (Underlining in the original)

Alhough one could ask how corporations almost *anywhere* would hold up to Swiss efficiency standards, even so, the picture here is bleak. Overemployment is, to some extent, a problem related to Nepal's powerful politically backed labor unions and labor laws that make it difficult to dismiss underperforming employees. But much more telling is the overall sense of malaise resulting from poor communication and a lack of employee investment or sense of responsibility to the larger enterprise. Reading this Swiss report, I could not help but think back to conversations with BPC employees from its early days when people remembered an invigorating work environment characterized by open communication, shared responsibility, creative problem-solving, and a meaningful sense of working for the common good. Now, whether due to passive inertia or active fear, BPC managers risked "no personal initiative," avoided "assuming responsibility," and spent their time "'waiting for instructions' from above"—the classic hallmarks of organizational paralysis. The Swiss consultant recommended that BPC implement a "**comprehensive turnaround plan** [*sic*]" that would require "strong" and "active" leadership from "**top management, the HR committee, and the management board** [*sic*]." From this report BPC appears as a rudderless ship adrift on a windless sea.

On the eve of its fiftieth anniversary (2015), BPC appeared to be in poor fiscal health, even though it consistently paid out sizable dividends. BPC's largest asset by far is its roughly 17 percent ownership share in Himal Power Limited (HPL), the owner–operator of the Khimti hydropower plant. Thanks to very favorable (and very controversial) terms in this pioneering plant's power purchase and other agreements, Khimti has been a huge profit source for its shareholders, including BPC. In 2015 BPC earned NPR 480 million (paid in USD) from its Khimti–HPL shares. That same year BPC paid out dividends to its own shareholders amounting to NPR 350 million—which means that BPC had to divert NPR 130 million of its Khimti earnings into its general operating budget just to break even. In short, without its Khimti–HPL earnings, BPC would have been insolvent and losing money fast.

CONCLUSION: BPC AT FIFTY YEARS AND BEYOND

After its privatization in 2003, the Butwal Power Company (BPC) had two primary goals: increase its power distribution territory and increase its company-owned generating capacity. In both of these areas the company's performance has been disappointing. One of the major components of Interkraft Nepal's (IKN's) privatization bid (and one of Hoftun's priorities) had been the aim of expanding BPC's existing distribution area (around its Andhi Khola power plant) to include multiple districts in western Nepal between Pokhara and Butwal. The hope was for BPC to grow into a major distribution company providing efficient electrical utility services directly to consumers. This did not happen. The government's vague promises to hand over Nepal Electricity Authority (NEA) distribution territory to BPC proved to be no match for NEA's unwillingness to give up any of its assets. It is also likely that the Shangri-La Energy Limited (SEL) investors did not share Hoftun's enthusiasm for expanding distribution which, especially in the short run, is never a money-making proposition.

Nevertheless, BPC *has* managed to significantly increase the number of consumers *within* its original distribution area. From only 2,000 or 3,000 customers, the privatized BPC has increased its consumer base to over 50,000, largely thanks to rural electrification infrastructure development grants from the United States Agency for International Development (USAID). BPC's distribution operations in the Andhi Khola area even retain some of the old "appropriate technology" elements from the original BPC–UMN integrated rural development (IRD) project. The old "cut-outs" have long since been replaced by meters but BPC still encourages consumer cooperatives which receive 10 percent discounts for taking care of bulk payments and routine repairs. BPC has emerged as a fairly major service provider, even if in a still relatively small distribution area.

Since its privatization BPC's total ownership of hydropower generating capacity has risen from 27.2 MW[20] to 33.9 MW,[21] a gain of only 6.7 MW. During the same period a host of other independent power producers (IPPs)—Nepali hydropower developers almost all of which were incorporated after 2003—brought 250 MW onto the market. Even the much-maligned NEA managed to add 50 MW of installed generating capacity to the national grid (Acharya 2017). A former BPC engineer noted that "not many megawatts have been added" but readily acknowledged that the blame did not fall entirely on BPC, even though it suffered from weak leadership. "The country's situation has made everything so difficult." But he had to admit

---

[20] Consisting of Andhi Khola, 5.1 MW; Jhimruk, 12 MW; and Khimti, 10.1 MW (16.88 percent of 60 MW).

[21] This figure includes the addition of Khudi, 2.4 MW (60 percent of 4 MW) and Andhi Khola "up-grade," 4.3 MW.

that, with other private hydropower developers moving aggressively into the market, BPC had lost a lot of its earlier status. "I think the pedestal that BPC was put on at the time of privatization—as a company with a lot of experience and depth: I don't think it was able to live up to that."

But BPC's long-term prospects are, perhaps, not as bleak as they seem. Even if they have not managed much by way of new installed generating capacity, since 2003 BPC executives and engineers have worked hard to advance several major new hydel projects which are now, at last, beginning to take shape. Planning and bidding for the 37.6 MW Kabeli A hydropower plant in far eastern Nepal, and the 30 MW Nyadi power plant in western Nepal, began soon after BPC's privatization in 2003. A third, much more ambitious, 100 MW Lower Manang Marshyangdi project was also on the drawing board. As of 2014 BPC had already made sizeable investments in planning and engineering for these projects: USD 4.6 million for Kabeli, USD 3.75 million for Nyadi, and a seemingly excessive USD 15.1 million in Lower Manang Marshyangdi.

But progress has been slow. As an example, a current BPC executive explained how work on Kabeli A began in 2005, "but then the whole process got so delayed because it's from the government." Even after winning a complicated international competitive bidding process, it took three years to sign a power development agreement (PDA) with the government and another five years to negotiate a power purchase agreement (PPA), gain environmental clearance, and negotiate financing. "So it took seven or eight years just to get through the bureaucratic hurdles." Again complicating matters was World Bank involvement: "You know the moment that happens, there's a lot of compliances to deal with."

This executive explained that lengthy delays represent not just a nuisance for the company but a threat to the very financial feasibility of the hydel projects themselves. As years stretch on, permits expire and financing terms become less favorable; transportation, equipment, materials, and labor costs rise; and a bid made almost a decade earlier becomes less and less feasible, let alone profitable.

> When we won the bid in 2007 the prices were one thing. But now it's 2016 and you can't do the project at the same rate. BPC will go ahead with

Kabeli but with the financing and increase in costs, it's not got going to be as profitable. If we go ahead on construction in the next few months, hopefully it can still survive.[22]

Financing these new projects has been a struggle, even with 40 percent of Kabeli funding coming from the World Bank or IFC. To finance the remaining portion of the USD 37.8 million project, BPC needed an international investment partner. One reason for the delay in securing financing for the Kabeli project was the question of majority control of the ensuing hydropower company's board of directors. At least two rounds of negotiations for Kabeli with international investors failed because BPC refused to concede board control—demanding at least 51 percent of company shares. In the end BPC did agree on financing with a Singapore-based investment group (InfraCoAsia), with the foreigners taking majority control. Interestingly BPC ended up with only a 40 percent shareholding stake in a joint-venture company with InfraCoAsia (Gurans Energy Limited) that, in turn, owns Kabeli Energy Limited, but BPC managed to negotiate ownership of 54 percent of the Kabeli project's output (20.3 out of a total 37.6 MW).[23]

Without World Bank involvement, the roughly USD 57 million, 30 MW Nyadi project has taken a very different route to financial closure. In this case almost all of the project's debt financing has come from Nepali banks and other financing institutions leaving BPC with a 60 percent share in the power plant's ownership and output. Without international investors—and the complicated, costly protective measures they often demand—Nyadi promises to be a more profitable (less cost per installed megawatt) project than Kabeli A.

Finally, in the spring of 2017 BPC broke ground on both the Kabeli A and Nyadi projects.[24] When these two projects go online BPC's ownership of installed peak generating capacity will have grown by 38.3 MW to 72.2 MW,

---

[22] Other industry insiders I spoke with had similar concerns about the Kabeli project's financial viability.

[23] From BPC board meeting minutes, document BPC BM 276-09.

[24] As I finalize this manuscript in June 2020, the prospects for Kabeli A again do not look good. The Chinese contractor hired by BPC has abandoned the works. And if NEA

an amount more than double BPC's current 33.9 MW capacity. If and when the major Lower Manang Marshyangdi project is up and running, BPC will have gained *at least* another 10 MW of owned capacity (and possibly more),[25] raising its total corporate holdings to 82.2 MW. When that happens, BPC will have (re)emerged as one of the major players in Nepal's independent power producing sector—though it remains to be seen how profitable these assets will be in terms of the cost–benefit ratios of BPC's investments.

---

goes ahead with its proposed Tamor reservoir project, the entire Kabeli A project will be inundated.

[25] On November 22, 2017, BPC issued a press release announcing a major new joint venture with a consortium of three Sichuan-based Chinese investment and engineering companies (both state owned and private). According to this announcement, the 100 MW Lower Manang Marshyangdi project will be the joint venture's first undertaking but there was no mention of what shareholding stake BPC will have in either the joint venture or the Lower Manang project specifically. The press release states: "The joint venture aims to develop at least 1,000 MW in next five years with investment of US$ 2–3 Billion." The consortium's leading (state-owned) investment firm has already developed and owns 37,000 MW of installed generating capacity in China. Whether BPC can hold its own in the company of these big hitters remains to be seen.

9

CONCLUSION

From Seed, to Plant, to Seed

> From my perspective, the most important story is BPC's impact on hydropower sector development in Nepal, *outside* of BPC itself.
>
> —Nepali hydropower executive, 2016

The Butwal Power Company's (BPC's) founder, Odd Hoftun, has long thought of his development efforts in Nepal as analogous to planting a seed, tending it with "faith and courage," and waiting to see how it might mature (Hoftun 2004). From his early days in Nepal, to his establishment of BPC in 1965, through the company's growth and development outlined in this book, Hoftun has watched BPC evolve in ways that are deeply satisfying in terms of his goals of human capacity building and promoting ethical business practices, but also increasingly personally dispiriting. As the seedling matured it increasingly took on qualities required of it by the soil and environment in which it had to grow. Despite his best efforts, Hoftun could only watch as his plant began to bear fruit that diverged from his ideals but reflected the demands of capitalist economics, global and local.

Symbolically, Hoftun's presence in the company also dwindled—from his early decades as general manager, through the disappointing privatization process where his hopes of securing board control for Interkraft Nepal (IKN) were reduced to a tiny 6.9 percent share of the newly privatized company after 2003. Although Hoftun had hoped to play a role in the new BPC, within a few years after privatization IKN's role dwindled to zero. From that point onward IKN could only wait and watch as its annual BPC dividends went to pay off its BPC debts. It took twelve years for the tiny, nonprofit IKN to clear its debts (with interest) before finally, in 2015, selling its BPC shares and severing ties completely. Hoftun had promoted an ethic of corporate nationalism—building hydropower infrastructure to strengthen Nepal's capacity at every level from rural electrification and urban industrialization to skilled manpower, to equipment production and construction skills, to national energy independence. As recently as 2006 Hoftun could write, "I have always looked at BPC as a small bulwark against international money interests"—interests that he saw as not just parasitic but actually predatory to Nepal's own interests. But by the time IKN sold its last BPC shares, the company was largely dancing to the tune of "international money interests."

So what had happened to Odd Hoftun's vision of building up Nepal's hydropower development capacity? Now in his nineties, Hoftun speaks of BPC with disappointment and, at times, bitterness. Judged by Hoftun's almost puritanical standards of ethical business practices, corporate reinvestment of profits, and alignment with the national good, BPC's performance is disappointing—as would be that of almost any other capitalist enterprise in Nepal or elsewhere.

But if we shift our perspective, broadening the focus to take in Nepal's hydropower sector as a whole, the picture becomes considerably more encouraging. Arguably Odd Hoftun—by focusing on the one "tree" that grew from the "seed" he planted more than fifty years ago—has overlooked the veritable *forest* of hydropower development activity that has sprung up as progeny of his initial efforts.

In this chapter I will argue—following the opinions of many Nepali industry insiders and observers—that Nepal's vigorous hydropower development scene today would have been virtually impossible without the pioneering efforts of Hoftun and countless other committed Nepalis and expats who grew BPC to maturity. The BPC story is far from the full story of hydropower history in Nepal: many other groups and individuals have also contributed. But without BPC's role in it, Nepal's hydropower experience would have been very different. Without BPC's contributions, Nepal's hydropower capacities would likely be similar to other "least developed countries" whose industries are almost completely dependent on foreign skills, labor, and supplies.[1] Beyond its own corporate contributions, BPC has long served as a kind of incubator, nurturing talent that is then disbursed into the wider world, where it too reproduces. The fact that Nepal today has hundreds of small- to medium-sized hydropower plants in operation, with hundreds more under construction and in planning—rather than a handful of mega power plants dropped onto the landscape by global "development" banks—is incredibly important. Even more important is the fact that *many* of Nepal's hydropower projects are being designed by Nepali engineers, built by Nepali civil contractors, equipped with major components designed and fabricated in Nepal, and funded by Nepali banks and other private investors in Nepali rupees (Karki 2017: 122). Nepal's sophisticated hydropower "ecology"—the complex, interdependent mix of designers, builders, suppliers, and investors—would not exist without BPC's pioneering role in setting that ecology in motion.

In this chapter I trace BPC's legacy in three interrelated domains: first, its contributions to corporate culture, hydropower legislation, and policy; second, BPC's role in promoting the development of human resources and industrial capacity in Nepal; and third, the question of national self-sufficiency in the hydropower sector.

BPC'S ROLE IN THE ESTABLISHMENT OF NEPAL'S INDEPENDENT POWER PRODUCTION SECTOR

It is easy to forget that, at its founding in 1965, the Butwal Power Company (BPC) was not only modern Nepal's first private hydropower developer but also among the country's first privately registered corporations.

---

[1] Perhaps most telling is the comparison between Nepal and the nearby Himalayan country of Bhutan. In the name of cooperation and aid, Bhutan allowed India to completely dominate its hydropower development, leading to a condition today of Bhutan's total dependence on India—a dependence that allows India to set increasingly extractive terms for asset development, power production, and sales (Saklani and Tortajada 2019).

Furthermore, BPC along with its offshoots (Himal Hydro, Nepal Hydro and Electric [NHE], Hydroconsult, Hydro Lab) together form the foundation that others have built on to form Nepal's complex hydropower development sector. In the 1960s Odd Hoftun chose the corporate model as his preferred means of developing hydropower in part because he wished to promote a values-based business ethic built around honesty, transparency, teamwork, corporate reinvestment, human capacity building, and national self-sufficiency.

In its early decades this corporate culture attracted scores of Nepalis who saw their own visions of national service, social justice, and hard work mirrored in Hoftun's corporate style. For this study I interviewed many of these early BPC (and subsidiary) employees. Almost all of them cited BPC's corporate culture as a source of inspiration and a guide to their own professional ethics even long after they had moved into other careers across the hydropower sector. But many of them also spoke of the difficulties of *passing on* those values to following generations without broader corporate support. Without institutional backing, individual managers or engineers cannot insist that junior colleagues resist practices common in the society around them. "That I cannot change," said one former BPC engineer. "The whole society should be changed, or even the whole world. That is the challenge. It's not a problem of Hoftun Sir's vision, but implementation is not possible." No doubt many Nepalis share some version of Hoftun's commitment to ethical business practices but living up to his lofty ideals in Nepal—or perhaps anywhere—may be incompatible with capitalism as we know it.

BPC spent most of its first three decades in the peculiar environment of Panchayat-era Nepal. With its quasi-planned economy, state-run industries, and largely agrarian society, Nepal's business environment for small, private entities was largely uncharted territory. Hoftun learned to operate in this make-it-up-as-you-go world by carefully managing the constraints of the Panchayat state. This meant inviting the state into BPC ventures as shareholders but never surrendering board control. In this way Hoftun secured the state's (often almost symbolic) investment in BPC and therefore cooperation, but without losing command over the company's priorities and practices.

Compared to the political and social turmoil of post-1990[2] Nepal, Hoftun looks back almost nostalgically on BPC's early years.

Before democracy came in, things were *predictable* in Nepal, *good and bad*! For our work, it was simpler to deal with government then. You know, a lot of development took place during those thirty years up to 1990. Since then has come lots of foreign money and involvement. But before, there was just the king who could say, "This is the situation" [and get things done].

Hoftun's point is not to celebrate the authoritarian past but to note a time when BPC's relations with the state were relatively manageable (often involving personal connections), and the government could still largely determine its own interests in the face of international pressures. After 1990 what Nepal gained in multiparty democratic freedoms they also lost in national self-determination with the state's coerced embrace of neoliberal globalization (Structural Adjustment, "free markets," foreign direct investment [FDI], and so on) (Rankin 2004).

In some ways the autocratic Panchayat regime served as a protective incubator for the fledgling BPC and its allied companies, offering them relatively benign regulatory neglect and shelter from foreign market competition. The BPC group of companies found their feet through their first three projects (Tinau, Andhi Khola, and Jhimruk). By contrast, Khimti no longer enjoyed the Panchayat era's protective cocoon and when the project encountered the post-1990 "real world" of international "open-market" investment and finance, it came as a staggering blow. To say that BPC and its corporate offspring had to "adapt" to new conditions misses most of the point. In fact BPC had to, along with the state and other interests, essentially *invent the new global capitalist conditions* under which private hydropower development could occur in Nepal.

Many of the hydropower experts I spoke with pointed to the new complex regulatory environment hammered out for the Khimti project as one of, if not *the* most important of, BPC's legacies for Nepal's hydropower sector.

---

[2] In 1990 began Nepal's first Jan Andolan, or People's Movement, the popular uprising that swept away the authoritarian Panchayat state that had been in place since 1961.

To facilitate Khimti, in 1992 the Nepal government implemented three new regulatory mechanisms (a Hydropower Development Policy, a Water Resources Act, and an Electricity Act) that, along with subsequent legislation specifically crafted to promote and protect Khimti's foreign investors, literally created the legal and regulatory environment that independent commercial power producers needed in order to exist in a new globalized market (Shrestha 2016). One industry insider argued that

> if BPC had not promoted Khimti to a level where it had reached the government, I think we'd have never gotten the [1992 Electricity] Act—the one that opened power production to the private sector.... If BPC hadn't promoted the project to that level, there really would have been no incentive for the government to actually come up with this legislation. It would have remained always government-owned projects. Who knows when we would have gotten that act?

In 1990 Nepal had two hydropower producer/developers. Between 1911 and 1990 the Nepal Electricity Authority (NEA)[3] developed 239 megawatts (MW) of hydel capacity while BPC had built (or was building) about 18 MW and owned 5.1 MW (Shrestha 2016). Though less than 1/10th the size of NEA, BPC's significance lay in its *very existence*: a small independent hydel development entity nestled incongruously within a massively state-dominated energy market. Because BPC had carved out a space for power production outside of state control, when the time came to open the market to private interests, the Nepal government already had a sense of what private development might look like and a partner it could work with. And even more importantly, when the necessary legal context was finally put in place in 1992, new independent developers entering the market could build on BPC's (and its spin-off companies') models and manpower. BPC had pioneered the ground that others would soon occupy and expand.

From *two* players (NEA and BPC) in 1990, by 1999 *twenty-two* private-sector hydel developers (independent power producers, or IPPs) had registered with the Nepal government of which eighteen were Nepali companies

---

[3] And its earlier permutations, see Shrestha (2016).

(including BPC). All of the new Nepali companies had registered since 1992 with most registering after the Khimti PPA was signed in 1996. And at least four of the seventeen new private hydel developers had management and/or top engineers who had previously worked for BPC (Pradhan 2000). After 1992 BPC entered the newly formalized private-sector development environment well ahead of the competition but already its seeds were sprouting on ground BPC itself had prepared.[4] By the end of 2017, hundreds of Nepali IPPs had, collectively, surpassed NEA in terms of total hydropower generating capacity (Acharya 2017; Karki 2017).

ARUN III

All of these post-1992 private-sector seedlings could have been smothered had Nepal accepted the controversial World Bank–backed Arun III hydropower project in the mid-1990s. The major (201 MW) Arun III project's demise is—along with other factors—another example of BPC's legacy. As discussed in Chapter 5, BPC's development logic—and its proven ability to build small- to medium-sized hydropower projects using primarily Nepali manpower and even equipment—inspired Nepali activists to push a development vision that stressed dozens or hundreds of small, locally planned and executed hydel projects rather than the large, dependency-inducing, foreign "aid" projects like Arun III (Pandey 1996, 2015).[5] In an interview twenty years later, Bikash Pandey explained how his experience working with BPC had equipped him with

---

[4] Of course it is ironic that, in the midst of this flurry of private entrepreneurial activity, in 1996 the Nepal government *nationalized* BPC, a move that both defied the neoliberal ideology of the time and significantly weakened BPC (Chapter 4). But the move also set the stage for another landmark achievement for BPC—the company's privatization, the largest and most complex in Nepal's history (Chapters 6 and 7).

[5] Amazingly, a clause in the World Bank's Arun III project agreement would have forbidden the government of Nepal and NEA from developing any hydropower project larger than 10 MW during Arun III's predicted decade-long construction period. Had Arun III gone ahead, NEA would not have developed its three largest hydel projects since 1995, or around 180 MW of currently installed capacity (Shrestha 2009).

an [anti-mega project] argument based on economics and development [rather than social or environmental concerns]. That was unusual [as a tactic for resisting World Bank impositions]. And one of the reasons you don't see many campaigns or arguments on that basis is that it's almost impossible to put the alternative out there to say *this would be better*. And the reason that there isn't the possibility of doing that is that, most countries haven't had an Odd Hoftun to have put [the alternative] out there already.

Pandey explained how big countries like India or China have political reasons and economic resources needed to develop their own hydropower industries. But most small, "least developed" countries have no choice but to accept the standard World Bank or Asian Development Bank (ADB) development logic which consists of foreign-designed and foreign-built projects dropped onto the landscape—projects that may generate power but not human capacity and self-sufficiency. Largely because of BPC, compared to many countries Nepal's indigenous hydropower sector is "very unique, very unique," said Pandey.[6]

> For a country at our level of development, it's highly unusual.... Because hydro requires all these elements: multi-year construction, local manufacturing, technical engineering, and construction. For that sectoral wide development [to be in place], well, it's pretty unique to Nepal.

Pandey admits that, when he and others made their case against the World Bank and Arun III in the mid-1990s based on the BPC experience, the evidence was promising but hardly comprised a solid statistical basis for projection. Based on their own experience, they knew that Nepali engineers, Nepali companies, and Nepali money *could* build high-quality, cost-effective power plants, but it was still a "big leap of faith in our minds." Speaking in 2016 Pandey noted that

---

[6] According to a US government website, while Nepal ranks 130th in the world in terms of electrical energy production, it is 10th in the world in its percentage of energy derived from hydroelectric sources (89.9 percent). See: "Nepal: Energy," The World Factbook, Cia.gov, https://www.cia.gov/the-world-factbook/countries/nepal/#energy (accessed January 4, 2018).

now, with twenty years of data, we can see a lot more points on the graph and they did indeed follow that curve [projected earlier]. The best companies follow it and, in aggregate, the whole sector follows it.... In any year there's anywhere from between 1,000 and 2,000 MW under construction and we should be out of load shedding by 2018 if not earlier.

Had Arun III gone ahead Nepal might now have one or several foreign-built mega projects but its load-shedding problems would likely be even worse than they are.[7] Because Arun III was stopped, and because the BPC model had paved the way for a small but diverse indigenous hydropower development sector, Nepal today has an independent national hydropower economy that is "very unique" for a developing country. Even if other comparable countries had promulgated the necessary legal environment to encourage private investment, they would not have had Nepal's already-in-place hydropower sector—thanks to BPC.

CAPACITY BUILDING AND MANPOWER DEVELOPMENT

Almost everyone I spoke with agreed that a crucial dimension of Odd Hoftun's vision for Nepal involved capacity building, technology transfer, and human development. When I asked Balaram Pradhan (who devoted much of his professional life to the Butwal Power Company [BPC]) what a book examining BPC's history should be about, he replied without hesitation: "BPC was established to institutionalize on-the-job training. How was that possible? *That is the story.*" He emphasized that BPC and its allied companies were founded not simply to build power plants, much less just make money, but to create skilled manpower. BPC's principle "product" was not hydropower but people with skills. Pradhan stressed that this focus

---

[7] Developing numerous small projects rather than a few huge ones also means a much more efficient use of dry-season water flows. Many small run-of-the-river power plants situated along a dry-season river can process the same small flow of water many times while a single mega project sits largely idle.

on human resource development (as opposed to infrastructure development) was often slow and inefficient. BPC's early power plants could have been built (by skilled outsiders) much more quickly and efficiently but those gains would have come at the cost of Nepali manpower competence and confidence building. Pradhan echoed the remarks of many other former BPC employees who stressed the value of "practical training": hands-on work that gave them the skills and—maybe even more importantly—the self-confidence to take on greater and greater responsibilities. "The point is not to say who did what in which year," said Pradhan. "That's not the story we need. But what was the basic vision of the company and what happened after that."

To illustrate his point Pradhan described joining a group of South Asian hydropower experts on an inspection tour of hydel plants in Bhutan, Nepal's Himalayan neighbor state. After visiting five huge, Indian-built power plants, the group sat down with Bhutanese officials.

> And they asked me, "What is the difference between what you saw in Bhutan and what you have seen in Nepal?" I said, "I have just one thing on my mind—that we have very similar resources and you are developing pretty fast while we are very slow. But that is not my concern—who is better or not. But I know one thing: training needs time. Good training needs *more* time. And we have trained more Nepalis in building a few hydropower plants. Here you have developed *a lot* of hydropower, but I can guarantee that none of the Bhutanese know how to build a hydropower plant. Who was trained here? Only the Indians. So what is the value of your power plants?

Pradhan challenged the Bhutanese officials to build even a single megawatt power plant without Indian help while noting that Nepalis are busy building their own hydroelectric future.[8]

---

[8] All of Pradhan's points are resoundingly confirmed by Saklani and Tortajada (2019) who outline the growing alarm within state and civil society circles in Bhutan over that country's total dependence on (and increasing exploitation by) Indian hydropower developers. See also Gyawali (2010).

It is difficult to quantify the manpower legacy of Hoftun's and BPC's work in Nepal but everyone I spoke with agreed that its impact on the hydropower sector is immense. Balaram Pradhan estimated that, in their first three decades (before BPC's nationalization), BPC and its spin-off companies trained around 5,000 people in high-value, practical skills ranging from engineering to accountancy, heavy equipment operation to complex civil construction, tunneling to high-tech welding, and quality control to project supervision, among many others. Pradhan challenged me to find, among the hundreds of Nepali companies now registered in the hydropower sector, more than a handful that were not employing BPC "family" graduates at one or more levels in their operations. Beyond BPC and its allied hydropower companies, other industry experts pointed to several consequential early Hoftun initiatives, such as the Butwal Technical Institute (BTI) (still going strong) and Butwal Engineering Works (BEW) (now closed), that have "led to this *huge* accumulation of manpower" in industrial manufacturing for the hydropower sector and beyond.

Today BPC is only one of "more than one hundred"[9] Nepali independent power producers (IPPs) and by no means the leader among them. According to one industry expert, in the years since its privatization, "BPC has hardly grown."

> The new owners have taken premiums out of what was set up as a pretty solid company with no debt. They've taken their dividends but have taken on debt instead of putting a lot of their own money into it. They haven't kept their leadership role.[10] But interestingly, in the meantime, the companies that used to be a tenth the size of BPC are now today the same size or bigger. In a sense, it was because BPC was there as an example, and also because there were competent people had worked for BPC before who left and joined these companies, that these companies really got a start. But

---

[9] According to Balaram Pradhan in 2015.

[10] Another industry analyst criticized BPC's new owners for focusing on short-term profits (on construction, supplies, equipment, and so on) rather than long-term energy investment. "These guys are not really energy investors. They end up making very expensive [unprofitable] hydro projects. The profit is not made as an energy company."

then those companies have really grown. I would say that that whole sector, without the example of BPC, wouldn't have been there, though it's always difficult to prove.

For him the fact that BPC is no longer an industry leader is far less important than that it served as a model and trailblazer for now more successful Nepali competitors.

If imitation is flattery, then BPC has much to be proud of given that many of its competitors are following BPC's lead beyond simply the role of hydropower developer. Like BPC, many have also set up in-house design and engineering departments analogous to BPC's Hydroconsult. And the same is true of BPC's corporate progeny: today the tunneling and general construction contractor Himal Hydro and the equipment manufacturer Nepal Hydro and Electric (NHE) both have numerous competitors, many with key staff members from the original companies. While NHE is still Nepal's industry leader (in terms of capacity, scale, and quality), the company struggles with smaller, leaner competitors who provide cheaper products to small-scale Nepali developers willing to sacrifice quality for cost (Shrestha 1995: 48). Many of these competing equipment suppliers are staffed by BTI graduates, and some have even established BTI-inspired in-house trainee apprenticeship programs. Similarly, today Himal Hydro has two major Nepali civil contracting competitors with all three thriving on demand for their services mainly coming from Nepali IPPs. The main point is that, across the hydropower development "ecosystem" (developers, engineers, contractors, and equipment suppliers), Nepal now has a proliferation of private players all competing in the space opened up by BPC and its allied companies.

How economically significant is Nepal's hydropower ecosystem? People I spoke with struggled to attach numbers to assessments of hydropower development's role in the national economy. One industry consultant estimated that there are "many tens of thousands of people" employed simply in construction at any one time in Nepal. Another current hydropower developer noted that if every project site had 1,000 people working (from professionals to day laborers), and if you included the food, transportation, and housing necessary for these people, plus the important related "upstream" (iron, cement, and so on) and "downstream" (transmission, and so on) industries, then "two or three hundred thousand very easily would be in this sector." Depending on how one estimates Nepal's labor force (a notoriously difficult proposition[11]), hydropower might employ up to 5 percent of the nation's workforce.[12] Although hardly precise, one industry analyst estimated that hydropower construction comprised "a not insignificant percentage of all private employment in Nepal."

Another way to think of the hydropower sector's impact is in terms of the money its development puts into the economy. Here again people often spoke of the legacy of Odd Hoftun and BPC. Surveying the expanding hydropower landscape, one former BPC engineer noted that, beyond producing generations of skilled Nepali manpower, a key part of Hoftun's legacy was a large and growing cadre of Nepali investors "trained in the art of taking risk." He recalled how, twenty years earlier, Nepalis had been

> scared to invest in 1 or 2 MW projects [but] now local investors are investing in 50–70 MW projects. And in the next ten years, I'm sure some of them will be doing 100 or 150 MW projects. So Hoftun is very much alive there.

The "art of taking risk" is not just about growing Nepali technical competence: it is about meticulously calculating relations of risk and reward and gradually increasing *confidence* in Nepali abilities. Looking back at Hoftun's career, this engineer noted that transferring the art of risk taking to Nepalis is not something that can happen in one person's career or even life span. "It takes fifty to one-hundred years. It's a slow-moving process."

One very significant dynamic driving Nepali investment in hydropower development is the simple fact that more and more Nepalis have more and

---

[11] The International Labor Organization estimates Nepal's workforce to be around 16 million ("Nepal: Labor Force, Total," Trading Economics, https://tradingeconomics.com/nepal/labor-force-total-wb-data.html [accessed November 30, 2021]) but conservative estimates place the number of Nepalis working abroad at 2.2 million while other estimates place the numbers at double that figure or more.

[12] Several Nepali hydroelectric industry experts noted that, in order to understand the unusual significance of the hydropower sector in Nepal, one only needed to look to the Himalayan regions east and west of Nepal where hydropower industries are miniscule.

more money to invest. Ironically, for one of the world's poorest countries, a small but growing number of Nepalis have money, especially Nepali rupees, to invest—with the bulk of the money coming from the country's now huge foreign remittance economy.[13] With relatives working in generally low-wage jobs abroad, few Nepali families have much surplus to invest. But in aggregate, the amount of available private investment capital is startling. A Nepali economist described how in Nepal, outside of the indigenous hydropower development sector, "there's no place to invest."

> There's one company, Hydropower Investment Development Company, which was set up to encourage investment in hydropower. So they invited people to buy their shares and they ended up being oversubscribed by thirty or forty times! Every time [private investors] are invited to buy shares in hydropower companies, there's a huge response.

Indirectly reflecting this same remittance-driven investment scenario, Nepali commercial banks and even public pension fund managers are mastering the art of risk-taking in the hydropower development sector (R. S. Shrestha et al. 2018: 9).

When I asked one hydropower developer to estimate the sector's contribution to Nepal's overall economy, he acknowledged that it was difficult to assess. He said,

> But if there's about 1,000 MW [of production capacity] under construction at any one time in Nepal, and if each megawatt costs about 1.5 million US dollars [to build], then you've got 1.5 *billion* US dollars being spent. That's in play now. That's going into the economy.

Some of that money is in foreign currency from foreign investors but much of it is in Nepali rupees from Nepali investors. "Then think about all the auxiliary businesses this money supports—food, housing, transportation,

---

[13] In 2013–14 Nepalis working abroad remitted over USD 5 billion (Sharma 2017: 55), or about 30 percent of Nepal's gross domestic product, almost five times the amount Nepal receives in international aid (Thompson, Gyawali, and Verweij 2017: 7).

upstream, downstream, you name it. Just consider the multiplier effect!" To the extent that it is Nepali investors funding Nepali developers employing Nepali engineers, contractors, and laborers, and purchasing Nepali-produced equipment, the recirculating economic impact of Nepal's indigenous hydropower sector exceeds any basic figures one can attach to it.[14]

NATIONAL SELF-SUFFICIENCY

The focus of this book has been the Butwal Power Company's (BPC's) role in promoting Nepali capacity and confidence in hydropower development. Based on Norway's experience, from the very beginning Odd Hoftun saw hydroelectric generating capacity to be the bedrock upon which Nepal could build a national economy. Without a cheap, sustainable, indigenous power supply—which, in Nepal, could only be hydropower—there could be no significant employment beyond subsistence agriculture, no modern industrial development, and no hope for any degree of Nepali national economic independence and dignity. While economic independence and self-sufficiency are always matters of degree, any hope that Nepal had of breaking from its historic role as an extraction zone (largely of human labor) lay in developing whatever economic strength it could muster to protect it from its giant neighbors, north and south. As described in earlier chapters, it was this pragmatic nationalist development ethic that attracted scores of like-minded Nepalis (and foreign volunteers) to the BPC cause. In the postcolonial era, Nepal's political independence would be in direct proportion to its economic—and energy—independence.

---

[14] Shrestha (2017: 143–147) contrasts the economic impact of Nepali-built and financed hydro projects, with donor-funded, internationally financed projects, arguing that the latter provide far fewer "backward" and "forward linkages" than the former. He cites the proposed 750 MW West Seti project to be built by an Australian developer with World Bank funds. Of an estimated USD 1.097 *billion* of total investment, only USD 28 million would be spent in Nepal (for labor, supplies, and so on). The rest would end up outside of Nepal to pay foreign contractors, debt financing, and loan insurance. Furthermore, because Nepal would own only 15 percent of the project, 85 percent of the plant's dividends would end up outside Nepal.

Yet if the Hoftun–BPC legacy is to be measured against Nepal's energy independence today, then it is—as yet—a legacy of failure. Although Nepal ranks an impressive tenth in the world in its percentage of electricity derived from hydel sources (89.9 percent), electrical power itself is still only a small part of the country's total energy consumption. Although it has among the world's highest *potentials* for electrical production, Nepal is still 130th in terms of actual electrical output.[15] Instead the large bulk of Nepal's energy consumption is in the form of imported fossil fuels and local biomass (wood, dung, grass).[16] In fact, Nepal is still desperately, chronically short of energy. Most of the country's hydroelectrical supply goes to simply keeping the lights on in urban areas,[17] with fossil fuels powering transportation and biomass still fueling cooking fires across rural Nepal. Nepal's industrial sector is still small and starved for power (though without the development initiatives described in this book, it would hardly exist at all). Hoftun's vision of cheap, abundant hydroelectricity powering a thriving industrial economy in Nepal is still many years away.[18]

The potential for realizing that vision remains: Nepal still has river resources capable of generating many tens of thousands of megawatts that could supply a booming industrial economy with power left over to export. The question is how best to move toward that vision of national energy

---

[15] See: "Nepal: Energy," The World Factbook.

[16] According to the government of Nepal statistics from 2017, in Nepal about 77 percent of energy used comes from biomass, 17 percent from fossil fuels (oil and coal), and only 3.32 percent from hydroelectricity (R. S. Shrestha et al. 2018: 5).

[17] Ironically, as consumers try to make up for power shortages, private diesel-driven electrical generators (in homes and businesses) now account for almost as much power production as Nepal's entire hydroelectric output (Thompson, Gyawali, and Verweij 2017: 7). Even more distressing is a recent turn to coal-fired production facilities, based on imported coal (P. Shrestha et al. 2018: 11).

[18] R. S. Shrestha et al. (2018: 6–9) argue that Nepal's installed hydropower generating capacity will need to rise from its current level of around 800 MW to at least 6,000 MW to meet domestic and industrial demand (current and expected) and to replace current usage of fossil fuels and biomass. Even now Nepal *imports* as much as 756 MW during dry-season peak demand periods, all of it from dirty coal-fired Indian power plants—and this only to keep the lights on in the major urban areas (P. Shrestha et al. 2018: 11).

independence. Should Nepal rely on its fast-growing indigenous hydropower development sector and continue to build small- to medium-sized projects around the country using mainly Nepali manpower, capital, and equipment? Or should Nepal turn to international hydropower developers (banks and private corporations), entities that have the expertise and capital to build massive power plants similar to the Arun III project that Nepali activists earlier rejected?

Almost every industry expert I spoke with in Nepal agreed that not only is this either/or opposition between national and foreign development a false choice but, in fact, the question has already been answered. Starting at Khimti (with the pioneering involvement of Statkraft), and accelerating in the last decade or so, Nepal has already gone far down the road of allowing large foreign developers (mainly Indian and Chinese) access to Nepal's hydropower development market. To my surprise, even the most ardent supporters of Hoftun's vision of a robust and self-sufficient Nepali hydropower development sector agreed that Nepal's demand for power is so great, and the need so urgent, that the answer is both: Nepal needs its own hydropower development apparatus working at full tilt alongside international developers. Or, as another Nepali developer put it, "Hoftun's vision is great but it's too slow for the needs of Nepal. Nepal isn't yet capable of doing big stuff, but it *needs* big stuff!"

Nepali engineers and developers I spoke with agreed that, as one person said,

> The bottom line is that we need large [internationally financed and built] projects. We [Nepalis] can build 10, 30, 50, even 80 MW plants. We need these and they make sense on our rivers. But we need much more. We need large storage projects [employing dams]. We need larger run-of-the-river projects. The national grid will only be viable if we have fairly large storage projects to meet peak demand.

Nepali developers now have considerable experience building small- to medium-sized run-of-the-river plants capable of generating a relatively steady base supply of power on any given day. The problem is that electricity demand is typically anything but steady. Especially when its primary consumers are urban households, electricity demand notoriously spikes in the morning and evening. For electrical power to be suitable for industrial use, the grid needs a reliably steady base supply (to keep machines running) along with the ability to generate peak surges to accommodate fluctuating household demand. The answer to this problem is storage which, in turn, requires dams: water stored behind dams is power that can be released onto the grid as needed to augment the continuous base supply.

If dams are the answer, they are also a big problem: storage facilities are technically challenging to build and, therefore, notoriously expensive. Perhaps especially in the Himalayas—where seasonal flow varies enormously, seismic dangers have to be accounted for, and siltation threatens to turn dam ponds into sediment traps[19] (and, therefore, eventually useless)— dams have to be massive and very carefully built. Because the cost of these ventures exceeds anything that Nepali investors can finance, storage projects inevitably require large-scale international financing. "And as soon as you get international financing, Nepali contractors don't stand a chance," one Nepali developer noted with a nod to BPC's own experience at Khimti (Chapter 5). He explained the catch-22 scenario whereby foreign banks consider it too risky to hire Nepali contractors to build projects larger or different than what they have already built, thereby making it impossible for Nepalis to gain the necessary experience and credentials.

> I mean that Khimti legacy still continues. If you want to do something really big, the funding required is not available in Nepal. And as long as the funding has to come from abroad, the funders will ensure that they have contractors and engineers and operators with international credibility.

Competition between Nepali and foreign developers is not a zero-sum game—there is enough urgent demand for all parties to prosper—but it is

---

[19] Odd Hoftun joked that poorly designed dam reservoirs in the Himalayas can end up so full of sediment that they are eventually nothing but "a nice flat area" in otherwise mountainous terrain, good for football but not much else! For example, the Kulekhani project's dam and storage reservoir were designed to function for 100 years but had entirely filled up after just 13 (Gyawali 2003: 131; Pandey 1997).

also not a risk-free scenario for Nepal. Even without the infamous under-the-table shenanigans that have plagued the Nepal government's dealings with mega foreign developers, how will the Nepali hydropower sector continue to expand its capacity (especially in design and construction experience) if they are systematically excluded from large-scale projects? One Nepali engineer described how developers like China's Three Gorges Corp. or the Indian GMR were heavily involved in major hydropower projects in Nepal but were typically employing foreign engineers, foreign construction companies, and sometimes even foreign basic labor. The kind of foreign-to-Nepali transfer of skills and technology that BPC and its collaborating companies benefited from at Khimti through its partnership with Statkraft (however problematically) are out of the question.

> Sure [foreign companies] have been bringing in projects within time and within cost, but there's no benefit for the country. It's just that they have built it. When they go away we are left without the capacity [in terms of new training and expertise]. *We have not built the capacity.*

Will Nepali developers ever be able to break through this scalar barrier?

Some of the Nepali developers and engineers I interviewed were confident in the Nepali hydropower sector's ability to grow on its own. They point to steady incremental growth in expertise and capacity over the past decades, as well as under-construction and proposed projects, and conclude that the future promises the same upward trajectory. One former BPC engineer noted that, even now, Nepalis are capable of designing major run-of-the-river projects of 1,000 MW and more.

But even the most optimistic Nepali industry boosters acknowledged two main interrelated obstacles to growth: financing and specifically dam-related expertise. Although it is wonderful that Nepali money is today funding many of the small- to medium-sized projects around the country, the fact is that available Nepali capital is miniscule compared with the country's demand for hydropower investment.[20] If Nepali developers could attract significant

---

[20] According to one Nepali industry analyst, in 2017 Nepali banks had a combined capacity to fund only about 50 MW of new hydropower construction *per year*. This

amounts of foreign money (foreign direct investment, or FDI), they could build bigger projects. But FDI is out of reach. Nepali developers repeatedly lamented about how international commercial lenders, *including* the World Bank, the Asian Development Bank (ADB), and other "development" lenders,

> now have no thought for development. All they care about is whether it gets finished and that their money is safe. They don't want to put any money aside for something like development experience or training and all that.

As had been the case at Khimti, still, whenever foreign money is involved, investment "safety" becomes such an obsession that virtually any Nepali involvement is either out of the question or such a potential security threat that it has to be heavily counterbalanced ("securitized") with layers of costly financial safeguards (Karki 2017: 118). Granted that capitalism and charity are rare bedfellows, but when even global "development banks" systematically refuse to promote Nepali development by refusing to include Nepali developers in their "development" projects, there seems to be little hope of Nepal ever gaining the proven track record need to attract FDI to its indigenous hydropower sector.[21]

If, as mentioned earlier, Nepal desperately needs large-scale dammed storage projects, and if these FDI-dependent mega projects systematically exclude Nepali engineers and contractors from participating, how will

---

represents a significant advance over recent years but it is still a drop in the bucket of the investment that Nepal needs.

[21] Karki (2017: 121) provides striking evidence of the cost-effectiveness of Nepali private-sector hydropower development as compared with large-scale, donor-financed projects. According to his analysis, in Nepal big public sector internationally financed projects average a cost of USD *3.623* million per installed megawatt (with the 70 MW Middle Marsyangdi project coming in at USD 5.714 million per installed megawatt); and international private sector projects financed by international commercial banks average USD *2.610* million per installed megawatt; while Nepali private sector projects with domestic financing in Nepali rupees average USD *1.832* million per installed megawatt. Nepal's private sector produces installed hydropower capacity at literally half the cost of World Bank and ADB financed projects, and it is Nepali consumers who pay the difference. See also Gyawali (2016).

the Nepali industry make the quantum leap into a large-scale future? The chairman of one of Nepal's leading civil contracting companies acknowledged that, while Nepalis can do a lot of things,

> we don't have the experience to construct high dams. For those things, these foreign companies, they have to come. And these big dams, they are not the projects to experiment with! You need *really good* companies to come and do them right the first time. But Nepalis, together with them, they can learn.

The question is how to make sure that Nepalis get together with *really good* foreign companies to learn the lessons of big dam design and construction. The first such partnership, between BPC and Statkraft at Khimti, was the result of a long history of relations between the companies and the fact that BPC had initiated the project. By the 1990s Statkraft wanted to take a more active role in Nepal's hydropower development and coordinating with BPC was essential in getting a foot in the door of the Nepal market. As pioneering multinational developers in Nepal, Statkraft *needed* partnership with a Nepali company in order to navigate the incredibly complex process of securing the conditions of international cooperation and foreign investment. But now, decades later, foreign development and investment proceed down well-established pathways such that, for foreign companies, partnering with Nepali companies seems unnecessary at best and disadvantageous at worst—if collaboration is seen as training potential competition.

GOVERNMENT SUPPORT

Nepali industry experts I spoke with agreed that the only way for Nepali companies to attract FDI, or for those companies to secure partnerships with more experienced foreign companies, was government intervention aimed at strengthening Nepal's indigenous hydropower sector.[22] Noting

---

[22] Karki (2017: 123) argues that, if anything, current Nepal government practices (if not policies) *weaken* the sector by favoring donor- and international-financed projects from which various taxes, fees, and bribes can be siphoned off.

that "Nepal tends to repel FDI," one hydropower booster argued that the Nepal government should offer foreign investors basic financial guarantees if and when they contract with Nepali companies. This would allow foreign investors (including banks) to limit their risks without having to add layers of very expensive international insurance that make hydro projects prohibitively expensive. In theory, government guarantees could both make FDI available to Nepali developers and keep the cost of construction per unit of power generation low enough to ensure a good likelihood of project profitability.

Other industry observers argued that the Nepal government should mandate that foreign development companies have Nepali partners—a relationship that, if properly managed, can be profitable for both sides. One former BPC engineer pointed to Khimti as an example of a successful, mutually beneficial joint venture. He estimated that 70–80 percent of the design and construction work at Khimti was done by Nepalis, but under the close guidance and supervision of Statkraft's Norwegian experts. Under this arrangement Nepalis learned skills ranging from heavy equipment operation to safety protocols, to technical design elements, to supply chain management. For its part, Statkraft acquired a major share in a profitable power plant with relatively little investment in manpower and resources. If project approval by the Nepal government was contingent upon joint venture relations such as this, it could be a win-win scenario with Nepali companies acquiring new capacities and foreign firms making profitable investments.

In fact the government of Nepal has, at times, instituted policies designed to promote joint ventures in hydropower development. An executive in one of Nepal's main construction companies pointed to the huge (456 MW) Upper Tamakoshi project as an example of both the potential for Nepal government's promotion of Nepali development interests and its pitfalls. He described how, initially, the government had put a clause in the Tamakoshi bidding contract offering a 5 percent price advantage to foreign developers who established joint ventures with Nepali partners. "So at that point [during the bidding process] international companies who were trying to bid, they were coming to us saying, 'Please become our partner.' And we were very happy!" Due to this government-mandated provision in the bidding procedures, the Nepali construction company actually had foreign developers competing for its collaboration.

> But suddenly—who is to blame we don't know—this clause was taken out. And then nobody cared for us. This Chinese company [Sinohydro] ultimately won the project. Earlier they came to us but when we approached them later, after the clause was removed, there was nothing. This thing is important. Nepalis have to learn working together with big international companies. Then very soon they will catch up. But it won't happen without government support.

The general assumption is that high-level Nepali politicians scuttled this pro-development clause in return for bribes from the foreign developers. The Nepal government could insist on policies that promote joint ventures, skills transfer, and capacity building in the indigenous hydropower sector. But placing national interests ahead of personal profiteering is not a reliable characteristic of Nepali political leadership.

Other industry observers called for even greater government intervention in hydropower development. One request was for a regulatory or planning commission that would try to rein in some of the hydropower sector's chaotic, market-oriented dynamics. For example, such a commission could try to promote a more equitable distribution of hydropower investment across Nepal. Rather than the current industry-driven scenario—which favors investment in projects close to established consumer markets—a hydropower development commission could prioritize investment in underserved areas (especially in western Nepal) to try to promote development outside of the core areas.[23] A commission could require that, in order to build a power plant near an urban area, a developer would have to build another project in a neglected area. As one industry veteran argued,

---

[23] R. S. Shrestha et al. (2018: 10) make a similar point: "Nepal needs a dedicated institution to prepare energy demand projections that aims to achieve self-reliant energy security with hydropower in every facet of household and industrial activities, and policy geared for government intervention to ensure the displacement of biomass and imported fossil fuel."

Some centralized planning authority should insist on production [in underserved areas]. The private sector and IPPN[24] isn't going to go there. The government is going to have to make these decisions. There should be somebody ordering these things, isn't it? That's why I say we need a regulatory body, *somebody* looking after the national interest.

Others argued that, in addition to trying to spread development more equitably across the country, a government-led hydropower commission could try to regulate *who* gets hydropower site development licenses and *how*. Returning to the theme of what is in Nepal's national interest, several observers complained of how a relatively small number of wealthy and politically well-connected Nepali business leaders have been able to snatch up licenses for the most promising sites, many of which then end up in Indian hands. "Who should be given these licenses?" one analyst asked.

How should they be given? Just because some rich IPP, someone with lots of family connections, goes to the ministry shouldn't mean that they get a license. *Somebody should be planning!* The same people are getting all the licenses, and then do you know what they're doing? They're selling them all to the Indians!

He estimated that about half of the 40,000 MW of generation licenses that the Nepal government has allocated, including most of the big projects, are now controlled by Indians. "Is that in our interest?" he asked.

Continuing the theme of calls for government involvement, several engineers I spoke with stressed the need for specifically hydropower-related technical training opportunities to be available *in Nepal*. They noted the irony of how, in a country with equally huge potential and demand for hydroelectricity, there is virtually no training available for hydropower engineers. One chief engineer (and former BPC employee) for a major developer complained of how the Nepali civil engineering graduates (of Tribhuvan University's engineering program) that he hires have very limited *theoretical* knowledge of

---

[24] The Independent Power Producers of Nepal (IPPN) is a trade association representing Nepal's hydropower developers.

issues specific to hydropower engineering, let alone any practical experience. He described how many of his junior engineers, after a few years of on-the-job training, go abroad for advanced technical training in various aspects of hydropower development—but then don't come back. "The problem is that we train, and then drain. Usually they go abroad. It is very difficult now in these days to retain people." According to him, the same pattern occurs across the hydropower development industry. "We have a high turnover of staff mainly because people want to study abroad and work there after gaining some initial experience in Nepal. It is very difficult to get them to come back."

One possible solution to this "train then drain" problem, he argued, would be for the Nepal government to create a "Hydropower University." Rather than having to enroll in specialized hydropower engineering programs in Norway, Russia, and elsewhere, the Nepal government could establish its own advanced training programs in Nepal, perhaps as part of Tribhuvan University's existing engineering department. Students could work in tandem with Nepali companies—perhaps even in some kind of public–private partnership—acquiring practical experience alongside advanced technical training. "Maybe this would help keep them in the country," he concluded somewhat doubtfully.

But his larger point was that Nepali private industry is not the ideal location for advanced training and national "capacity building." Inevitably skills have to be honed in on-the-job settings but most corporate enterprises are not interested in building an overt educational function into their overall business models. Corporate success might be good for the national interest, but the inverse is not necessarily true. Realistically, he suggested, businesses should not be expected to sacrifice self-interest for the national good. There may be times when profit-building and nation-building can go hand in hand but in the end private capitalist interests will choose the former over the latter. For better or worse, he said, the nation should not look to private industry to automatically advance the national interest. Across the board, he concluded,

> For Nepali capacity to really grow, Nepal's hydropower development sector needs government support. For that there should be development policy. Unfortunately, my feeling is that we don't have this willingness or feeling to develop institutions in Nepal. Even the government is not supporting.

CONCLUSION

I could not help but notice how many of these calls for governmental intervention in Nepal's hydropower sector—along with their misgivings—resonate strongly with the most basic premises of Odd Hoftun's development philosophy, values that had been embedded in the original Butwal Power Company's (BPC's) corporate ethos. The irony is that the set of corporate values that BPC stood for prior to its nationalization and then privatization are now seen as so threatened that state intervention seems like the only solution. The priorities that a committed team of Nepalis and expats had pursued under a formally private corporate structure during the Panchayat era served well both the company and the country. BPC grew, its skills expanded, and—together with its corporate offspring—it established the complex corporate ecology (of development, design, civil construction, and equipment manufacturing) from which sprang Nepal's current dynamic hydropower development sector.

But neither BPC nor its original values have fared well under the post-1990 global free-trade dispensation. While a privatized BPC has struggled to stay afloat (due to debatable management priorities) in a competitive scene that now includes powerful multinational players, the market as a whole has not proven very hospitable to the ethical vision to which Odd Hoftun, Balaram Pradhan, and many other loyal BPC staffers (Nepalis and expats) had committed their careers. Odd Hoftun had wanted to prepare BPC—and Nepal's hydropower development industry broadly—for the harsh realities of market competition. That indigenous industry is now vigorous and promising, even if much of the corporate vision that BPC stood for has seemingly fallen victim to market forces. With the market unable or unwilling to sustain these values, concerned industry insiders turn to the Nepal government as the last hope of keeping them alive.

Perhaps foremost among these early BPC core values are a set of related concerns surrounding Nepali capacity building; prioritizing the national interest; national self-sufficiency; and profitability through efficiency and reliance on local skills, resources, and labor. Balaram Pradhan had pointed to human resource development as BPC's principle legacy. Even if it was slow and problematic from a strictly economic point of view, providing on-the-job training was, he believed, "the story" when it came to assessing BPC's impact on Nepal. Although a private corporate entity, BPC placed the interests of Nepal's national development (including the needs of the rural poor) above that of its corporate bottom line. Hoftun and Pradhan saw profits as essential to corporate sustainability but understood the role of profit not as a means of enriching management or shareholders, but of growing the company and advancing its abilities to train manpower and promote national development. In turn, profitability would come not from cutting corners or exploiting labor, but from pursuing maximum efficiency by relying on skilled Nepali labor, local resources, and appropriate technology whenever possible.

If today's private, market-driven hydropower sector is unable or unwilling to uphold these values, it is somewhat ironic that industry insiders (including former BPCers) would turn to the Nepal government to keep them alive—given that one of Odd Hoftun's most quixotic battles has been with the forces of corruption in both Nepal's economic and political sectors. From the earliest days Hoftun mobilized BPC's collective forces around a shared commitment to corporate financial transparency, social responsibility, and ethical business practices in both its relations to other businesses and the government, even if it meant swimming against the heavy tide of standard Nepali business and political practice. Recognizing the Nepal government as among the chief arbiters of BPC's fate, as well as a powerful and potentially negative force, Hoftun had sought to minimize predatory political and bureaucratic interventions by the government in BPC's affairs by including the state among BPC's official owners, but strictly as minority shareholders without board control. This approach, though not without risks, was the most likely to secure governmental cooperation while minimizing governmental interference. In fact, as I described in Chapter 1, Hoftun had chosen the private corporate model as a means of distancing his project from the pitfalls of state involvement. That so many of those who share Hoftun's ethical vision would now turn to the state to uphold it suggests that the private, commercial, corporate structure was, ultimately, unable to sustain that vision.

With industry experts calling for government interventions ranging from foreign direct investment (FDI) protections and mandated joint ventures to regulatory bodies that would guarantee social justice and state investments in manpower training, the question becomes to what extent the BPC model is, or was, a viable platform from which to pursue these goals outside of the state's purview. The BPC model of ethical practice could fit into a corporate business structure as long as those with board control remained firmly committed to those values. But to what extent are corporate values like these compatible with capitalism, and especially the cutthroat world of international market forces and competition? As this study has shown, Hoftun's values came under sustained attack as soon as BPC was exposed to global market forces—at Khimti. As Statkraft shifted the basis of its relations with BPC from altruism to profit maximization, and as international business financing came into play, BPC managed to stay afloat, but only after having compromised some of its values, or simply had them overridden. What was left of those values in BPC's corporate culture dwindled further through the traumatic experiences of nationalization, privatization, and BPC's reorientation under Nepali ownership.

In Chapter 1 I described how, at its inception, BPC had three different institutional logics woven into its "corporate DNA": a service and social justice logic represented by the United Mission to Nepal's (UMN's) participation and partial ownership; a nation-building agenda represented by the Nepal government's co-ownership; and a market-based, commercial business orientation institutionalized in BPC's official structuring as a for-profit corporation. I noted how Hoftun's somewhat unique background as a Norwegian Christian allowed him to sustain a corporate vision that held these three logics—social justice, nation building, and profit—in creative tension. Hoftun then spent decades struggling to prove that "ethical capitalism" was not the oxymoron that his critics believed it to be. BPC was founded on a belief that a corporate entity *could*—with enough discipline and enlightened leadership—earn profits in order to promote social advancement and serve the national good through capacity building and energy independence. For decades Hoftun and others fought to preserve a fragile bond between ethics and capitalism within BPC.

This study has traced the gradual fraying of that bond—within BPC and the hydropower development sector in Nepal as a whole. Social justice, national interest, and profit could be harnessed into one agenda, but only as long as corporate leadership dedicated to that vision remained strong. Hoftun had chosen the corporate model as the platform from which to pursue his vision precisely because he understood that dedicated leadership—in the form of board control—was the only way to guarantee its implementation. If the early chapters of this book traced BPC's struggles to knit together a set of somewhat incompatible, contradictory priorities, the later chapters (from Chapter 5 onward) documented their slow unraveling as those committed to BPC's founding vision gradually lost control of, and even influence on, BPC's leadership.

Today, the various parts of the original BPC vision live on, though in separate pieces. BPC's dream of developing Nepali human capacity is embodied in a vibrant indigenous hydropower development economy that is virtually unprecedented in the "developing world." BPC's role in nurturing this sector is by far its most important legacy. But across the industry the private corporate *will to prioritize capacity building as an end in itself*, not just a means to profitability, is gone. Likewise, many hydropower industry experts and advocates (including many now working in the commercial hydropower sector) still embrace BPC's early commitments to serving the disadvantaged and serving the national interest. But all of them have given up on Nepali industry's ability and/or willingness to pursue ethical business, much less the public good. As a last resort, they turn to the state as the only means to mandate ethical behavior through regulation, even while recognizing that the state itself can hardly be counted on to pursue the nation's best interests.

BIBLIOGRAPHY

Acharya, Pushpa Raj. 2017. "IPPs Set to Overtake NEA in Power Production by Year-End." *Himalayan Times*, February 25, 2017. https://thehimalayantimes.com/nepal/independent-power-producers-ipps-set-overtake-nepal-electricity-authority-nea-power-generation-year-end. Accessed December 8, 2021.

Bakkevig, Ludvig Johan, Odd Hoftun, and Hallvard Stensby. 1996. *Jhimruk Hydro Electric and Rural Electrification Project in Nepal: Experiences from the Project Implementation*. Oslo: Norwegian Agency for Development Cooperation (Norad).

Bhandari, Vinay, and Gyanendra Lal Pradhan. 2006. "Andhikhola: A Model for Sustainable Rural Development." *Hydropower and Dams* 1: 78–80.

Bhattarai, Binod. 2001. "BPC's Privatization Saga." *Nepali Times* 57, August 24–30, 2001. http://archive.nepalitimes.com/news.php?id=9220. Accessed December 8, 2021.

Bishwakarma, Meg B. 2007. "Addressing Sediment Problems." *International Water Power and Dam Construction*, May 15, 2007. http://www.waterpowermagazine.com/features/featureaddressing-sediment-problems/. Accessed December 8, 2021.

Butler, Christopher. 2014. "May You Live in Interesting Times." *Hydro Nepal* 15 (July): 21–22.

Chaudhary, Binod. 2015. *My Story: From the Streets of Kathmandu to a Billion Dollar Empire*. Kathmandu: Nepa-laya.

Dahl, Gjert Aaberge. 2014. "Hydraulic Design of a Francis Turbine That Will Be Influenced by Sediment Erosion." MA thesis, Norwegian University of Science and Technology, Department of Energy and Process Engineering, Trondheim.

Dhakal, Resham Raj. 2004. "Construction Ropeways in Nepal." In *Ropeways in Nepal: Context, Constraints, and Co-evolution*, edited by Dipak Gyawali,

Ajaya Dixit, and Madhukar Upadhya, 85–96. Kathmandu: Nepal Water Conservation Foundation.

Gyawali, Dipak. 2003. *Rivers, Technology, and Society: Learning the Lessons of Water Management in Nepal*. Kathmandu: Himal Books.

——. 2010. "The Neocolonial Path to Power." *Himal South Asian*, August 1, 2010. https://www.himalmag.com/the-neocolonial-path-to-power/. Accessed December 8, 2021.

——. 2015. *Nexus Governance: Harnessing Contending Forces at Work*. Gland, Switzerland: IUCN. DOI: 10.2305/IUCN.CH.2015.NEX.5.en. https://portals.iucn.org/library/sites/library/files/documents/Nexus-002.pdf. Accessed December 8, 2021.

——. 2016. "Grand Larceny: Reverse Robin Hood behind the Privatization of Infrastructure." *Spotlight* (Kathmandu) 10 (7). https://www.spotlightnepal.com/2016/11/17/grand-larceny-reverse-robin-hood-behind-the-privatization-of-infrastructure/. Accessed December 8, 2021.

Gyawali, Dipak, Ajaya Dixit, and Madhukar Upadhya, eds. 2004. *Ropeways in Nepal: Context, Constraints, and Co-evolution*. Kathmandu: Nepal Water Conservation Foundation.

Gyawali, Dipak, Michael Thompson, and Marco Verweij, eds. 2017. *Aid, Technology, and Development: The Lessons from Nepal*. London; New York: Routledge.

Hoftun, Odd. 1997. "Kathmandu Water Supply Scheme: A Three Component Package Proposal with Special Reference to the Melamchi River Diversion Scheme, November 1997." Unpublished bound mimeographed report.

——. 2004. "Butwal Power Company Ltd.: Its Past, Present, and Future." Unpublished manuscript.

——. 2007. "History and Recent Developments of Melamchi Diversion Scheme." Unpublished manuscript, dated December 10.

——. 2011. "Interview with Odd Hoftun." *Hydro Nepal* 8 (January): 81–83.

——. 2013. "Norwegian Involvement in the Melamchi Project." Unpublished manuscript, dated June.

Horst, Otho. 2018. *Eight Little Words: How God Led a Mennonite Farm Boy to a Remote Town in Nepal*. Bealeton, VA: Apples of Gold Media.

Inversin, Allen R. 1994. *New Designs for Rural Electrification: Private Sector Experiences in Nepal*. Arlington, VA: National Rural Electric Cooperative Association.

Jantzen, Dan, Bikash Pandey, Ratna Sansar Shrestha, and Odd Hoftun. 2008. "Multi-Purpose Melamchi Project." *Hydro Nepal* 8 (July): 35–37.

Karki, Ajoy. 2017. "Micro and Small Hydro: Serial Leapfrogging to a Braver New Nepal." In *Aid, Technology, and Development: The Lessons from Nepal*, edited by Dipak Gyawali, Michael Thompson, and Marco Verweij, 113–131. London; New York: Routledge.

Kshetry, Jiwan. 2018. "The Lokman Saga: How Citizens, Parliament, and the Judiciary Challenged the Abuse of Nepal's Anti-corruption Watchdog by Its Chief." *Studies in Nepali History and Society* 23 (1): 173–214.

Kunwar, Shyam Bahadur. 2016. "Resistance and Negotiation: An Ethnographic Reflection of Jhimruk Hydropower Project (JHP), Western Nepal." Unpublished manuscript.

Møgedal, Tor. 1983. "Some Old Ideas, Some New Ideas, Then Lots of Determination and Hard Work: The Story of Butwal Technical Institute." Unpublished manuscript.

Nafziger, Dale. 1990. "Impacts and Implications of Rural Electrification Ideology in Nepal's Domestic Sector." PhD dissertation, Cornell University.

New Business Age (Kathmandu). 2016. "CSR Initiatives of Some Companies in Nepal." December 19, 2016. http://www.newbusinessage.com/MagazineArticles/view/1665. Accessed December 8, 2021.

Onta, Pratyoush. 1996. "Creating a Brave Nepali Nation in British India: The Rhetoric of Jāti Improvement, Rediscovery of Bhanubhakta, and the Writing of Bīr History." *Studies in Nepali History and Society* 1 (1): 37–76.

Pandey, Bikash. 1996. "Local Benefits from Hydro Development." *Studies in Nepali History and Society* 1 (2): 313–344.

———. 2015. *Track II: Nepal's Pluralistic Hydropower Development Post-Arun*. Chautari Foundation Lecture 2015. Kathmandu: Martin Chautari Press.

Pandey, Kumar. 1997. "Attitudes Towards Development: A Look Back at Kulekhani." https://groups.google.com/forum/#!topic/soc.culture.nepal/mhbrA93qaMg. Accessed December 8, 2021.

Pokhrel, Anil. 2017. "Water Supply and Sanitation: Elusive Targets and Slippery Means." In *Aid, Technology, and Development: The Lessons from Nepal*, edited by Dipak Gyawali, Michael Thompson, and Marco Verweij, 167–184. London; New York: Routledge.

Poppe, Joy. 1993. "Andhi Khola Project, United Mission to Nepal, Socio-Economic Survey, 1993." Unpublished report, United Mission to Nepal, Kathmandu.

Pradhan, Balaram. 2000. "Private Sector Companies for Hydro Power Development in Nepal: Some Information." Unpublished manuscript, dated January.

Pradhan, Pratik Man Singh. 2004. "Improving Sediment Handling in the Himalayas." *OSH Research, Nepal* (October): 1–6.

Pun, Santa Bahadur. 2010. "The Controversial Khimti PPA: The Roles of HMG/N's EDC and International Lenders." Unpublished manuscript.

Rankin, Katharine N. 2004. *The Cultural Politics of Markets: Economic Liberalization and Social Change in Nepal*. Toronto: University of Toronto Press.

Saklani, Udisha, and Cecilia Tortajada. 2019. "India's Development Cooperation in Bhutan's Hydropower Sector: Concerns and Public Perceptions." *Water Alternatives* 12 (2): 734–759.

Sharma, Sudhindra. 2017. "Trickle to Torrent to Irrelevance? Six Decades of Foreign Aid in Nepal." In *Aid, Technology, and Development: The Lessons from Nepal*, edited by Dipak Gyawali, Michael Thompson, and Marco Verweij, 54–74. London; New York: Routledge.

Shrestha, Nawa Raj. 1995. "The Role of Nepal Hydro and Electric and Butwal Engineering Works in Developing Hydropower Equipment." In *Report on a National Seminar on Mini- and Micro-Hydropower Development in the Hindu Kush–Himalayan Region*, edited by R. D. Joshi and V. B. Amatya, 45–48. Kathmandu: ICIMD.

Shrestha, Pranay, Durga Prasad Upadhyay, Shankar Prasad Sharma, and Dipak Gyawali. 2018. *State of Climate Action in Nepal: Nepal-CANSA Annual Snapshot, 2018*. Kathmandu: Nepal Climate Action Network South Asia (N-CANSA).

Shrestha, Ratna Sansar. 2009. "Arun III Project: Nepal's Electricity Crisis and Its Role in Current Load Shedding and Potential Role Ten Years Hence." *Hydro Nepal* 4 (January): 38–43.

———. 2011. "Our Opposition Is Only against Colonization of Nepal's Natural Resources." June 28, 2011. http://www.ratnasansar.com/2011/06/our-opposition-is-only-against.html. Accessed December 8, 2021.

———. 2015a. "FDI in Hydropower and Choice of Jurisdiction." *Hydro Nepal* 16 (January): 33–34.

———. 2015b. "Do Your Writings Bring About Any Change?" March 9, 2015. http://www.ratnasansar.com/2015/03/do-your-writings-bring-about-any-change.html. Accessed December 8, 2021.

———. 2016. "Hydropower Development: Before and after 1992." April 2, 2016. http://www.ratnasansar.com/2016/04. Accessed December 8, 2021.

———. 2017. "Large Hydro: Failures in Financial Engineering." In *Aid, Technology, and Development: The Lessons from Nepal*, edited by Dipak Gyawali, Michael Thompson, and Marco Verweij, 132–152. London; New York: Routledge.

Shrestha, Ratna Sansar, Stephen Biggs, Scott Justice, and Amanda Manandhar Gurung. 2018. "A Power Paradox: Growth of the Hydro Sector in Nepal." *Hydro Nepal* 23 (July): 5–21.

Shrestha, Tara Lal. V.s. 2074. *Bandipurdeki Bandipursamma*. Kathmandu: Discourse Publications.

Svalheim, Peter. 2015. *Power for Nepal: Odd Hoftun and the History of Hydropower Development*. Kathmandu: Martin Chautari.

Thompson, Michael, Dipak Gyawali, and Marco Verweij. 2017. "The Dharma of Development." In *Aid, Technology, and Development: The Lessons from Nepal*, edited by Dipak Gyawali, Michael Thompson, and Marco Verweij, 3–12. London; New York: Routledge.

Upadhya, Madhukar. 2017. "Bhattedanda Milkway: Why a Climate- and Mountain-Friendly Technology Continues to Be Ignored." In *Aid, Technology, and Development: The Lessons from Nepal*, edited by Dipak Gyawali, Michael Thompson, and Marco Verweij, 77–94. London; New York: Routledge.

Veltmeyer, Henry, and Paul Bowles, eds. 2018. *The Essential Guide to Critical Development Studies*. New York: Routledge.

Veltmeyer, Henry, and Raúl Delgado Wise. 2018. *Critical Development Studies: An Introduction*. Halifax, Nova Scotia: Fernwood Publishing.

Weller, G. A., Tor Skeie, and Dan Spare. 1992. "The Andhi Khola Project, Nepal: Hydropower in Development." In *Hydropower '92*, edited by E. Broch and D. K. Lysne, 727–733. Rotterdam: A. A. Balkema.

INDEX

ABB Energi, 69, 88, 95, 127
Acharya, Mahesh, 188
Acharya, Shailaja, 118n3
Adhikari, Purna Prasad (P. P.), 124, 149
African Ministerial Conference on Hydropower and Sustainable Development, vi, 255, 255n9
Agder Energi, 199–201, 207n8, 209, 212, 217, 226, 231–32, 239–40
Alstom, 245–49, 251, 254
Andhi Khola hydropower project, 9, 21, 24, 28, 51, 59–85, 88, 88n3, 89, 91–93, 96–102, 115–16, 119, 131, 137, 165, 167, 171, 180n16, 197, 211, 214, 231–32, 252, 265–66, 266nn20–21, 274; "upgrading" of, 232, 252–58, 260–61, 261n15, 262–63
Andhi Khola Water Users Association (AKWUA), 78–80, 82n20
appropriate technology, 8, 26, 43n, 67, 74, 75n17, 76–77, 92, 254, 266, 296
Arun III (World Bank–funded) hydel project, 130, 130n8, 131, 131n9, 132, 132n10, 276, 276n5, 277–78, 286

Asian Development Bank, 121, 128–30, 133, 152–53, 162–63, 179, 191–92, 196, 277, 289, 289n21

Bahadur, Dhan, 34
Basnet, Suman, 227, 231, 233, 239, 239n2, 242–43
Bhattarai, Baburam, 4n6
Bhattarai, Krishna Prasad, 186, 188
Bhutan, 22, 272n1, 279, 279n8
Birendra, King, 50, 144n10
Blue Planet Prize, 76n18, 254–55, 260
Bockmann, Kristian, 34
Butwal (city), 7, 28–29, 31, 40, 45, 48, 54–55, 57
Butwal Engineering Works (BEW), 53–53, 60, 68, 280
Butwal Power Company (BPC), ix, xi, xin1, 8–12, 17–28, 33–34, 37, 39, 41–43, 45–46, 49–52, 54, 54n14, 57–58, 64–65, 65n5, 67–68, 68n11, 69, 69n12, 70–71, 73, 73n16, 74–77, 79–84, 86–87, 87n2, 88–90, 94–96, 98–99, 99n8, 100–04, 106–07, 109, 109n16, 110–11, 111n17, 112–31, 134–40,

INDEX

142, 144–59, 159*n*7, 161, 161*n*9, 162–71, 171*n*13, 172–80, 180*n*16, 181–202, 204, 206, 206*n*6, 208, 208*n*9, 209, 211, 212*n*10, 213, 215–21, 223, 223*n*17, 224–50, 250*n*6, 251, 251*n*7, 252–61, 261*nn*16–17, 262–68, 268*n*24, 269, 269*n*25, 270–76, 276*n*4, 277–82, 284–85, 288, 290–91, 293, 295–298; founding of, 33–34, 272; Kathmandu headquarters controversy, 260–63; nationalization of, 9–10, 77, 106–12, 126, 145, 148, 150, 276*n*4, 295, 297; shell company, as, 51–52, 67. *See also* privatization (of BPC)

Butwal Technical Institute, x, 7–8, 15, 28–29, 31–33, 35–61, 66–67, 67*n*9, 85, 103, 109, 136, 280–81; trainee program, 35–39, 47

capacity building, x, 6–9, 11–12, 15, 23–29, 40–41, 46–48, 60, 126, 130*n*8, 137, 143–45, 148, 163, 195, 227, 230, 242, 245, 255, 270–73, 277–84, 292, 294–95, 297, 298

Chaudhary, Binod, 174–75, 204, 207, 211–13, 215–16

Chaudhary Group, 159*n*7, 173–75, 177–79, 181–82, 185, 188, 195, 201, 204–06, 206*n*7, 207, 207*n*8, 210–12

Chhetri, Rudra Bahadur, 37

Clarke, Douglass, 153, 176, 184, 189–90, 196, 203, 206, 208

"corruption." *See* Hoftun, Odd: "corruption," critique of

Critical Development Studies, x

dams and storage facilities, 50, 71, 90–91, 101, 136, 286–87, 287*n*19, 288–90

Development and Consulting Services (DCS), 7*n*9, 43, 43*n*, 50, 54, 57*n*15, 58–59, 68, 68*n*11, 75–77, 131

development, critique of, x, 8, 12–13

development philosophy, x–xi, 5–6, 8, 10–27, 112, 114, 120, 124–27, 145, 163, 170, 189, 203, 230–31, 237–38, 241–43, 246, 248, 252, 254–55, 264–65, 269, 271–73, 276, 295–98

East–West Highway, 7, 28, 31, 53
education versus training, 7, 46–50
Electricity Act (1992), 118, 197, 275
Electrification, rural, 61–63, 73–77, 80, 93, 99, 131, 138, 266, 271; urban, 43, 45, 50–51, 285, 285*n*17, 287
Environmental Impact Assessment (EIA), 34, 66, 97, 99, 101

foreign aid and investment, xi, xi*n*1, 10, 12, 12*n*11, 18, 26, 30, 32, 33*n*2, 34, 52–53, 58, 60–62, 64, 67, 69, 79, 81, 84, 86–87, 90, 92, 103, 106–07, 113–14, 116, 119, 124, 151–52, 155, 168, 179, 192, 225, 272*n*, 276, 283*n*13, 284*n*14, 289*n*21, 290–91, 296

Foreign Investment and Technology Transfer Act, 127

Galyang, 59, 64–65, 65*n*6, 71, 77, 82
Ghamire, Laxman Prasad, 133*n*11
government. *See* Nepal, government of
Grierson Report, 81
Gugeler, Martin, 29–30
Gyanendra, King, 220*n*15

INDEX

Gyawali, Deepak, 130, 222, 222*n*16, 223, 223*n*17, 224

Hagen, Egil, 227, 231–34, 239–40, 242–44, 253
Harwood, Peter, 123, 123*n*6, 124
Hegglid, Gunne, 172–73, 185*n*19
Himal Hydro and General Construction Company (Himal Hydro), 42–43, 52–53, 53*n*11, 54, 55, 58–59, 65–66, 68–69, 71, 81, 84–87, 89, 91–95, 112, 115–16, 119–20, 125–27, 134–37, 141–43, 146, 149–50, 175, 183, 259, 273, 281
Himal Power Limited (HPL), 117, 119, 121–24, 127–29, 132–33, 133*n*11, 134–35, 139, 142, 144, 199*n*1, 206, 209, 265
Hoftun, Martin, 119
Hoftun, Odd, ix–xi, xi*n*1, 2–4, 4*n*6, 5–22, 26–60, 60*n*1, 61–62, 62*n*2, 63–64, 64*n*4, 66–69, 76, 80–87, 88*n*3, 89, 91–92, 94–95, 95*n*6, 98–100, 107–09, 109*n*16, 110, 112–17, 119–24, 127–29, 136, 142–45, 147–56, 158, 158*n*6, 160–67, 169–71, 173, 173*n*14, 174–80, 182–84, 186–90, 192, 194–201, 203–06, 208–12, 212*n*10, 213–16, 216*n*13, 217–22, 225–41, 244–46, 246*n*4, 247–48, 248*n*5, 249–50, 250*n*6, 251–55, 257–58, 258*n*13, 259–60, 263*n*18, 270–71, 273–74, 277, 280, 282, 284–85, 287*n*19, 295–97: "corruption," critique of, 12, 16–17, 20–21, 38–39, 178–79, 185*n*19, 232, 234, 245, 248, 248*n*5, 250, 255–56, 262, 273, 288, 292, 296; early life, 2–4; management style, 20–22, 26, 41, 44; missionary colleagues' criticism of, 11, 16–17, 28, 48, 99–100, 100*n*10, 109–10, 115–16; religious, political, and ethical convictions, 3–4, 12–17, 19–20, 26, 36*n*4, 37–38, 40, 297
Hoftun, Tullis, 4–5, 7*n*8
Hydro Consult, 42–43, 58, 68, 68*n*11, 69, 85, 89–91, 99, 111, 116, 120, 149–50, 152, 191, 273, 281
Hydro Lab Private Limited, 91, 105, 135
Hydro Nepal (journal), 260
Hydro Solutions Limited, 260
Hydropower Investment and Development Company, 283
hydropower regulation and legislation. *See* Nepal, Government of

Independent Power Corporation (IPC), 168, 173, 175, 177–79, 181–83, 185–86, 188, 204–05, 206*n*7, 207, 207*n*8, 210
Independent Power Corporation Nepal (IPCN), 210–13
industrialization, ix, 1, 7–9, 11, 21, 28, 32–33, 38, 43, 45–48, 50–51, 54, 56–57, 60–63, 68*n*11, 69, 80–81, 83, 103, 187, 271–72, 280, 284–85, 285*n*18, 287, 292*n*23
InfraCoAsia, 268
Integrated Rural Development (IRD), 64, 80, 82, 98, 266
Interkraft (Norway), 156–58, 169, 172–73, 200
Interkraft Holding Limited, 202
Interkraft Nepal (IKN), 155–58, 160, 159n7, 164–65, 168–73, 175–80, 180*n*16, 181–85, 185*n*19, 186–92, 198–200, 200*n*2, 201–02, 202*n*3,

203–07, 207n8, 208, 210–14, 214n12, 215–22, 225–28, 230–46, 249–50, 252–53, 258, 270–71
International Finance Corporation (IFC), 121, 129, 133, 256, 268
International Hydropower Association (of UNESCO), 254
Institute of Technology and Industrial Development (ITID), 7, 7n9, 29, 43–44, 54
Inversin, Allen, 75
irrigation, 21, 23, 34, 63–64, 64n4, 66, 75, 77–79, 82n20, 97–98, 102, 138, 181

Jan Andolan ("People's Movement" of 1990), 96, 100, 106, 113, 115, 274n2
Jantzen, Dan, 110, 166, 169, 173, 180–81, 184, 186, 188, 233, 239n2
Japan International Cooperation Agency (JICA), 192
Jhimruk hydropower project, 9, 80, 83–112, 115–16, 119, 134, 136–40, 143–45, 150, 164–65, 167, 171, 180n16, 183, 198, 208–09, 214–16, 218, 220, 223, 223n17, 224, 224n18, 226, 238, 241, 252, 258, 266n20, 274; bombing and repair of, 102, 214–16, 220, 223–24, 232, 244–52, 262; community relations/mitigation problems, 96–103; siltation and erosion problems, 103–06
Jhimruk Industrial Development Center Pvt. Ltd. (JIDC), 103
Jhimruk Rehabilitation Project (JRP), 245, 249, 251–52, 258, 262
Jyoti, Padma, 209, 213, 218–19, 222, 225, 235, 237, 243, 247, 249, 251

Kabeli A hydel project, 267, 268, 268n22, 24
Kaligandaki River, 51, 59, 64, 66, 71, 77–78
Karki, Lokman Singh, 206, 211, 215
Karki, R. B., 244
Khimti Environment and Community (KEC), 139
Khimti hydropower project, 9, 18, 53n, 111n, 112–47, 149–51, 154, 158, 165, 182–83, 192, 199n1, 201, 206, 209, 241, 265, 266n20, 274–76, 286–91, 297; construction, 133–38; financial closure, 127–33, 150; power purchase agreement (PPA) controversy, 122, 128–29, 132–33, 133n12, 158, 182, 199n1, 209, 265, 276. See also Himal Power Limited (HPL)
Khudi hydel project, 259, 258n13, 259–61
Khudi Hydropower Limited (KHL), 259
Kirne, 119, 134–35, 139
Koirala, Girija Prasad (G. P.), 152, 185–86
Kunwar, Shyam Bahadur, 98, 102–03

labor, export of, 1–2, 146–47, 193, 282n11, 283, 283n13, 284, 294; training of skilled, 6–9, 11, 15, 23–24, 28–29, 36–37, 40–41, 46–48, 57, 67, 85–86, 89–90, 94–95, 139, 144, 170, 242, 254, 271, 278–80, 288–89, 293, 296
Lockwood, Peter, 137–38
Lower Manang Marshyangdi hydel project, 267, 269, 269n25

INDEX

Mahat, Ram Sharon, 153–54
Maoist conflict in Nepal. *See* People's War
market and market conditions, creation of, ix–x, 2, 10–13, 13*n*, 14–15, 17–19, 26, 31, 38, 43, 45, 48, 51–52, 57–58, 62, 65, 69, 82, 86, 90, 106, 107*n*15, 113–14, 117, 120, 145–46, 156, 171*n*13, 230, 246, 266–67, 272–75, 281, 286, 290, 295–97
Martin Chautari, 119*n*4
Marwari, 159*n*, 173–75, 201
Melamchi Diversion Scheme (MDS), 143, 146–94, 203; Multi-Purpose Melamchi Project (MPMP), 180, 181*n*17; "three component package proposal," 152–54
Melamchi Water Limited, 150
mitigation, 96–103, 138–40
"multiplier effect," development and, 8, 284

Nepal Electricity Authority (NEA), 54, 54*n*14, 55, 65, 67, 73, 73*n*16, 76, 84, 96, 104, 107–08, 110–11, 115, 118, 118*n*3, 123*n*6, 128–29, 132, 133*n*11, 145, 145*n*19, 161, 164, 167, 171*n*13, 176, 182, 184, 191, 194, 196–97, 204, 206*n*6, 208*n*9, 212*n*10, 223, 223*n*17, 225, 256, 266, 268*n*24, 275, 276, 276*n*5
Nepal Hydro and Electric (NHE), 42, 58, 58, 68–69, 69*n*9, 71, 73, 81, 84, 86–88, 88*n*3, 89, 93–95, 95*n*6, 105, 112, 114–16, 127, 134, 136–37, 143–44, 146, 150, 238, 245, 250, 250*n*6, 253–54, 261*n*16, 273, 281
Nepal, government of, 7, 11–12, 17, 23–24, 26, 29, 33*n*2, 37, 43, 45–55, 57, 64–66, 73, 79, 84, 87*n*2, 95–96, 98–102, 106–11, 113–18, 121, 123, 126–27, 130, 132, 144–45, 148–54, 159, 161, 163, 169–71, 171*n*13, 172–73, 175, 177–80, 180*n*16, 181–99, 203, 206, 209–10, 212*n*10, 213–27, 232, 235, 245–46, 250*n*60, 258, 262, 267, 274–75, 276*nn*4–5, 285*n*16, 290–96; as corporate minority shareholder, 8, 10, 12, 18, 32–33, 50, 53, 65, 92, 96, 108–10, 159, 163, 225, 232, 273, 296; hydropower regulation and legislation, 34, 44*n*7, 54*n*14, 65*n*, 66, 97, 113–14, 118–19, 127, 144, 145, 145*n*19, 154, 197, 217, 250, 272, 274–75, 292–94, 296; Panchayat era (1961–90), 96, 100, 102, 115, 273–74, 274*n*2, 295
Nepal Industrial Development Corporation, 32–33, 33*n*2, 225
Nepali Congress (party), 115, 133*n*11, 185–86, 189, 211
New ERA (consulting service), 49
Newar, 159–60, 204
Nippon Koi, 153, 191
Norad, 18, 30–32, 33*n*2, 52–53, 58, 62, 64–65, 69, 86–87, 90–92, 97, 103, 106, 115, 117, 129, 139, 151, 154–58, 162–63, 165–66, 168, 180, 180*n*16, 181, 183, 186, 188–90, 192, 211–12, 217, 220–22, 224–25, 239, 247–48
Norfund, 199, 218, 226
Norway, 2, 2*n*2–3, 3–5, 8, 12, 20, 24, 30, 32, 34, 36, 42, 47, 51, 53, 54*n*13, 61, 66–67, 67*n*9, 68, 74, 76, 81, 84, 87–89, 90*n*5, 94–95, 106, 108, 117–18, 120, 125, 129, 142, 144, 148, 151–52, 156, 161–62, 164, 169, 172–73, 186–88, 197, 200, 202,

205, 209–10, 212, 217–19, 221–22, 225–27, 231, 237–38, 240, 245, 249, 254, 294
Norwegian National Institute of Technology (Trondheim), 54, 90
Nyadi hydel project, 267–68

organic analogy, development and, vi, 9, 33, 58, 60n1, 255, 269, 272, 276
Organization of Petroleum Exporting Countries (OPEC), 192

Panchayat era. *See* Nepal, government of
Pandey, Bikash, 130–31, 186, 276–77
Pandey, Simon, 42
Pax men, 6, 29–30, 45n10
People's Movement (of 1990). *See* Jan Andolan
People's War (Maoist conflict in Nepal, 1996–2006), 102, 111, 135n, 140, 145, 154, 162, 182, 192, 194, 198, 203, 206, 211, 214–15, 244, 261
Pokhara, 7, 28, 51, 54, 59, 62, 73, 266
Poppe, Duane, 65, 65n6, 66–67, 79
Poppe, Joy, 65, 67
Pradhan, Balaram, 4n6, 20, 22, 26, 44–15, 45n10, 46, 51, 53–54, 58, 64, 82, 89, 107–08, 110, 151, 158–61, 166–69, 171, 182, 185, 190–91, 194–96, 199, 204, 208, 210–11, 213, 215, 218–19, 220, 224, 227–28, 231–33, 247, 249, 278, 280, 280n9, 295
Pradhan, B. K., 115
Pradhan, Bharat B., 32
Pradhan, Gyanendra Lal, 158–61, 167, 178, 185, 202, 208, 210–12, 215–16, 216n13, 218–19, 222, 227, 233–37, 239–40, 243–44, 246–47, 249, 252, 254–56, 258–60, 263
Pradhan, Pratik Man Singh, 136–37
privatization (of BPC), 9–10, 110, 148–229, 245, 269, 276n4, 295, 297; first bid, 161, 164, 166–75, 183; second bid, 183–91; third bid, 195–208; fourth bid, 208–27
Privatization Cell, 153, 163–69, 171, 175–77, 179, 182, 184–85, 188–89, 196, 203, 207–10, 213, 214n12, 227. *See also* Clarke, Douglass
Pun, Santa Bahadur, 133n11
Pyuthan District, 83, 100, 102, 214

"quality auditor," 87, 89

Rajbhandari, Rajiv, 218, 236–38, 242, 247
Rana, D. R., 39
Ravnevand, Tor Einar, 241, 243–44
rock and rock conditions, 5, 32, 34, 53, 69–71, 91–93, 136, 141, 141n16, 142, 166, 191, 193, 256, 258n13
Rokaya, K. B., 235–36, 238, 241
ropeways, 75, 75n, 144, 146, 165

safety protocols, 34, 70–71, 137–38, 291
Shangri-La Energy Limited (SEL) 158–61, 173, 175, 201–05, 209–11, 215–16, 216n13, 217–20, 220n14, 221–22, 225–28, 230–31, 233–37, 239–40, 240n3, 241, 243–48, 248n5, 249–50, 250n6, 251, 251n7, 252, 254–55, 257–62, 266
Sharma, R. P., 39
Sharma, Shiva Kumar, 69, 137, 142, 149

Shrestha, Badri Prasad, 223–24
Shrestha, Dil Bahadur, 136
Shrestha, Ratna Sansar, 82, 118, 169, 186, 218, 222, 233–38, 240, 240n3, 241, 243–44, 246, 249–50, 253–54, 257n11, 259
Shrestha, Tara Lal, 37n5, 44n8, 158–59, 159n7
SKK Energi, 172, 199, 205
Sorumsand/Kvaerner, 2n4, 66, 66n8, 67n, 69, 88, 94–95, 106, 127–28, 143–144
Statkraft, 117–24, 127–29, 132–34, 136–39, 141–43, 172, 183, 189, 199n1, 286, 288, 290–91, 297
Stiftelsen, 200, 200n2
structural adjustment/neoliberal "reform," 107, 113, 119–21, 153, 173, 274, 295
Swisscontact Senior Expert Corps, 263, 263n19, 264

Tansen, 5–8, 12–13, 15, 28–29, 51, 59
technology transfer, 8, 67–68, 88, 111, 127, 143–44, 170, 200, 240, 245, 254, 278, 288, 291–92
Tinau hydropower project, 8–9, 28, 31–35, 39, 41–43, 45, 50–54, 54n14, 55, 59–61, 65–66, 71, 83, 96, 98, 106–07, 274
Tribhuvan University Institute of Engineering, 90, 105, 135, 275, 293–94
Trisuli hydel project, 256
tunnels and tunneling, 24–25, 34–35, 42, 50, 53, 59, 62n2, 71, 84, 90, 92–93, 119, 134–35, 137, 140–41, 148–50, 152, 154, 163, 165–66, 169–80, 180n16, 181, 181n17, 183, 189, 191–92, 256

turbines, 2n4, 32, 42, 50, 55, 66–67, 67n9, 69, 73, 86, 88–90, 94–95, 95n6, 97, 104–06, 138, 143, 150, 198, 209, 252–54, 256; Francis, 90, 94, 105, 105n14; Pelton, 66, 73, 105, 105n14, 143, 253

United Nations Development Program (UNDP), 149–50
Unified Marxist Leninist (UML) party, 4n6, 110, 132
United Mission to Nepal (UMN), 4–8, 10–11, 16–17, 20, 24–25, 28–29, 33, 33n2, 34, 37, 41–43, 48–50, 52, 53n12, 54, 58–66, 68, 68n10–11, 70, 78–82, 82n20, 83, 85–87, 87n2, 89, 91, 99–100, 103, 107–09, 109n16, 110–12, 115–17, 126, 136–37, 139, 143, 149, 166, 173n14, 225, 262, 266; disagreement among expats, 40–41, 48: expats and volunteers, 19–20, 22, 24–25, 28–29, 30, 37, 39–42, 49–50, 58, 67–68, 68n10, 70, 80–81, 85, 89, 91–92, 100, 111–12, 116, 136, 139, 271, 284, 295. *See also* Hoftun, Odd: missionary colleagues' criticism of
Upadhyay, Dinesh, 42
Upper Tamakoshi hydel project, 140n15, 291
used, second-hand, or refurbished equipment, 30, 32, 40, 43, 50, 62, 62n2, 66–69, 73, 76, 79, 88n3, 92, 140, 208n9, 252–53

"Washington Consensus," 106–07, 130

Wasserkraft Volk (WKV), 228–29, 247–48, 250–51
water, as a resource, 1–3, 60, 97, 101, 103, 162, 285
Water Resources Act, 118, 275
Winrock International, 242

World Bank, 121–22, 127, 129–30, 130n8, 131, 149, 152–53, 179, 196, 256–57, 267–68, 276–77, 284n14, 289, 289n21

Yami, Hisila, 4n6